CRYSTAL CLEAR

Harnessing the Combined Potential Power of Crystals and Water

by

Noelle Sherrard

Copyright © 2011 Noelle Sherrard

All rights reserved. No part of this book may be used or reproduced by any means, graphic, electronic, or mechanical, including photocopying, recording, taping or by any information storage retrieval system without the written permission of the publisher except in the case of brief quotations embodied in critical articles and reviews.

Balboa Press books may be ordered through booksellers or by contacting:

Balboa Press
A Division of Hay House
1663 Liberty Drive
Bloomington, IN 47403
www.balboapress.com
1-(877) 407-4847

ISBN: 978-1-4525-4296-6

Library of Congress Control Number: 2011961616

Because of the dynamic nature of the Internet, any web addresses or links contained in this book may have changed since publication and may no longer be valid. The views expressed in this work are solely those of the author and do not necessarily reflect the views of the publisher, and the publisher hereby disclaims any responsibility for them.

The use of crystals has not been medically evaluated. The author of this book does not dispense medical advice or prescribe the use of any technique as a form of treatment for physical, emotional, or medical problems without the advice of a physician, either directly or indirectly. The intent of the author is only to offer information of a general nature to help you in your quest for emotional and spiritual well-being. In the event you use any of the information in this book for yourself, which is your constitutional right, the author and the publisher assume no responsibility for your actions.

The views expressed are not necessarily those of the Publisher.

Any people depicted in stock imagery provided by Thinkstock are models, and such images are being used for illustrative purposes only. Certain stock imagery © Thinkstock.

Printed in the United States of America

Balboa Press rev. date: 03/08/2012

This book is dedicated to Margaret who re-activated my interest in crystals and to all the other Aboriginal people associated with Marr-Mooditj Health Foundation who helped me to understand the "love of country". Thank you.

"We shall not cease from exploration, and the end of all our exploring will be to arrive where we started and know the place for the first time."

From *Little Gidding*
T.S. Eliot

Acknowledgments

The journey down the path of writing and creativity is new to me. What a challenge, with numerous by-ways to explore which opened up broader vistas. People who have accompanied me on this journey have all willingly contributed vital support across a wide ranging spectrum – Thank you all.

A journey always starts at the beginning hence my acknowledgements are made in that order. Margaret Quartermaine, friend and mentor, who re-ignited my interest in crystals; Sue Bussau, my niece, who assured me I wasn't too old for such a journey! Helen Quinn for encouragement, fiery enthusiasm and assistance with computer and camera; Stan at The Crystal Cavern for knowledge willingly shared and assistance with sourcing crystals; Jarrad Lenegan, my nephew, who no doubt despaired as he tried to teach me the very basics of chemistry and who was my guide dog as he led me gently around the Table of Elements; Suzanne Dewar whose active participation, contribution to photography and consummate skill and painstaking attention to detail on all matters relating to the computer – my heartfelt thanks; Margaret and Robert Barnacle for wise counsel and encouragement over numerous lunches; Maggie Taylor who maintained a watchful eye on my health during some tough times; Angie (Divine Crystals) for willing support, encouragement and artistic approach to her photography; Rosemary Hood for sound and valuable advice on matters pertaining to crystals and energy; Laurie Maddison for professional expertise in the critiquing of this book; Maureen Soklich for sound and practical advice regarding the crystals and where to source them; and last but by no means least, Barry, Stan's friend, whose knowledge of crystals and their chemical components is worthy of high praise.

Table of Contents

ACKNOWLEDGMENTS ... vii
INTRODUCTION ... 1
PROLOGUE ... 2
HOW TO USE THIS BOOK ... 3
ENERGY MEDICINE ... 4
CHILDREN OF THE EARTH ... 6
CREATION ... 7
WATER ... 9
MAGNETISM ... 13
FARMING PRACTICES AND SOIL ... 14
ROCKS ... 15
THE CELL ... 17
CRYSTALS ... 21
    CRYSTAL SYSTEMS ... 21
    HEXAGONAL CRYSTALS AND SHAPES WITHIN NATURE ... 22
    HYPOTHESIS ... 23
    NUMEROLOGY AND CRYSTALS ... 23
    THE DINSHAH CONNECTION ... 25
    EMPIRICAL AND RECORDED KNOWLEDGE OF CRYSTAL USES ... 26
        ATLANTIS ... 26
        AYURVEDA ... 26
        AUSTRALIAN ABORIGINES ... 27
        SOUTH AMERICAN INDIANS ... 28
        TIBET ... 29
        HILDEGARD OF BINGEN ... 29
MINERALS ... 31
    ELEMENTAL COMPOSITION OF THE HUMAN BODY ... 31
    MINERALS - HELPFUL INFORMATION ... 33
METAPHYSICAL AND OTHER USES OF CRYSTALS ... 39
MINERAL TINCTURES ... 50
    PREPARATION OF MINERAL TONICS ... 51
    DOSAGES ... 52

APPENDIX 1 – THE CRYSTALS CHOSEN 55
APPENDIX 2 – THE PICTURES. 56
APPENDIX 3 – TABLE OF ELEMENTS 71
APPENDIX 4 – CHEMICAL CONTENT OF THE 52 GEMS/MINERALS 75
APPENDIX 5 – CHEMICAL CONTENT - ADDITIONAL HEXAGONAL & TRIGONAL/
HEXAGONAL GEMS/MINERALS. 78
APPENDIX 6 – WHERE TO FIND THE MINERALS 80
REFERENCES 85
BIBLIOGRAPHY. 87
INDEX 89

INTRODUCTION

Information presented in this book has been compiled because my concern is about the direction in which our health, as a nation, is heading.

Primarily my interest is with health and clearly other members of the global community also are looking to the past for solutions to, and reasons for, the dilemma of life in the twenty-first century. This is evidenced by renewed interest in the Mayan civilisation and the lost continents of Atlantis and Lemuria. With potentially insurmountable problems of global warming, financial meltdown, rising unemployment, diminishing oil supplies, extremes of wealth and poverty, overburdened health systems, drug resistant super-bugs, constant wars, civil unrest, an increasing refugee population and last but by no means least, the present instability of the planet itself, humanity faces the most fundamental question of all - how will we survive?

During my twenty plus years as a naturopath and herbalist it became clear to me that many illnesses in our society are the result of poor nutrition. It was not that sick people were hungry, it was because there was a lack of nutritionally sound food in their diet. That conclusion led me to ask "why?". The answer to that question is to be found in the following chapters.

I hope this book will challenge and then encourage some readers to seek alternative ways to attain the basic requirements for healthy life – good water, essential mineral intake and thus nutrition.

The use of crystals in attaining this healthy life may be considered unconventional by some in modern society but I believe this method to have been used by indigenous people, to good effect for a long period of time, who believe that what is contained in the Earth is contained within us.

As conventional/allopathic medicine faces the burden of greater financial costs, over-crowded hospitals, chronic illness and the use of drugs that are no longer effective, there may be a need to acknowledge that alternative methods of health care will be required. What is presented herein may be viewed by some readers with scepticism, as it is scientifically unproven. However future developments could well lead to acknowledgement that alternative therapies should be sought.

PROLOGUE

Without a physical body that is both functional and healthy we cannot properly access our non-physical selves. Outlined in the following chapters are basic requirements needed by our body to allow this progression to occur. Our physical body is the vehicle we use for transport on our Life journey. Without a regular supply of good quality nutrients the human body is unable to perform effectively. Good quality food and water are fundamental to all life; minerals are also key elements in the food chain. Without them we would be unable to survive. Likewise water; the human body consists of approximately seventy per cent water; it is essential for our survival.

Alternative ways to access these key elements are fundamental to our health and well-being. Sections in this book make that clear on current availability of minerals and water.

The human body has its own innate intelligence. This book provides information on how to access that intelligence to incorporate the physical need for minerals at an energetic level.

Many have assumed that current food and water supplies are health-giving and adequate for their needs. Some are finding this is not so. Health practitioners are now of the opinion that ninety per cent of illness emanates from either lack of or imbalance of minerals. As we continue to exploit and vandalise Planet Earth, showing little or no respect to Mother Nature, we ourselves are beginning to experience health problems on an unprecedented level. The present health system in most western countries is based on the concept of closing the stable door after the horse has bolted. Costs related to these health systems are neither fully effective nor viable and will continue to increase exponentially.

I contend it is time to leave behind the notion of separateness and embrace the reality that we are all one. What we do to Planet Earth we do to ourselves! Mother Nature has provided us with everything we need to maintain a healthy and fulfilled life but many, as the result of outside influences, have become confused.

Maintaining health was not meant to be complicated. Retracing our steps to some of the old ways and accompanying wisdom would be a place at which to start!

HOW TO USE THIS BOOK

This book may be read as one from beginning to end, or randomly read as each chapter is complete within itself.

Information contained herein supports my theory concerning the use of crystals/minerals as a source of mineral supplementation.

The appendices provide details of the various aspects of the crystals including crystal type, individual chemical content, total chemical content and the Table of Elements.

Photographs of the sixty-nine crystals are for identification purposes and/or accessing one's personal requirements through divination.

ENERGY MEDICINE

When I commenced studying with AMORC (Ancient & Mystical Order of the Rose Cross) Organisation some 30 years ago, the concept of everything having its own rate of vibration was not much more than a theory. To a large portion of the population this was a fanciful notion and not worthy of investigation. Consequently these teachings were not publicly proclaimed. It was necessary to become a member of an organisation to obtain access to the knowledge. A few years prior Russia was engaged in research relating to psychic phenomena. A book on the findings *PSI – Psychic Discoveries Behind the Iron Curtain* by Sheila Ostrander and Lyn Schroeder was published in Great Britain by Sphere Books Ltd in 1973.

After reading excerpts from the book it became clear to me that Russia exhibited an accessible un-mystical approach to the mystical and unseen world. References were made and explanations given on the work of Semyon Davidovich Kirlian and his wife, Valentina. Kirlian photography today is accepted as part of the world of energy. The Rosicrucians, in their teachings, talked of two bodies – the physical body and the non-physical body. The Kirlians had made it possible to view not only the physical body, but also the non-physical body, that is, the energy body.

Through continuous research and application with plant material the Kirlians were able to identify illness in plants before it was visible to the naked eye; it seemed as though the physical body replicated the energy body; if there appeared an imbalance or distortion in the energy body, gradually this change would manifest in the physical body and the plant would die.

This discovery gave credence to modalities like Homoeopathy, Flower Essences and Acupuncture in which the energy body or non-physical body is the body being treated. It was postulated that the root of all illness lay in the energy body and if allowed to continue residing there would materialise in the physical body.

Today this view of the Energetic Medicines has become widely accepted amongst non-allopathic practitioners. What is of interest to me, as a practitioner and someone who is drawn to the use of minerals in a curative sense, is the concept of using crystals (which contain minerals) as energy medicine rather than minerals in their gross material state. (Energy medicine being non-physical, gross material being the physical state.)

Gem Essences are used extensively by non-allopathic practitioners. In the main, their usage, however, is confined to correction of emotional issues. Homoeopathy, on the other hand, includes numerous minerals in its arsenal of remedies as a means of addressing problems which have already manifested in the physical body. An example of this is Mercury (Mercurius) - "Every organ and tissue of the body is more or less affected by this powerful drug; it transforms healthy cells into inflamed and necrotic wrecks, decomposes the blood, producing a profound anaemia. This malignant medicinal force is converted into useful life saving and life preserving service if employed homoeopathically, guided by its clear cut symptoms." This description on the use of mercury is the preamble on Mercurius taken from Boericke's "Materia Medica" (1987, p. 432).

The difference between a homoeopathically prepared substance and a gem essence is quite specific. Homoeopathically prepared substances are based on several components - the law of let like be cured with like and the law of the minimum dose as well as the process of trituration and potentisation. Making gem essences, by comparison, is a simple process. The vibration of the mineral/crystal is unique and is held in liquid form and summing up - crystal, water and time are all that is required. Alcohol may be added to increase the shelf life. (See *Mineral Tinctures*.)

As well as treating the non-physical body, homoeopathy treats the illness once it has manifested in the physical body. What I am trying to determine is whether there is a therapeutic value in Gem Essences based on their energetic mineral component. Could this energetic mineral component act as a prophylactic for the physical body?

Through the late nineteen eighties and early nineteen nineties when magnetism was being investigated in the West as a therapeutic tool, Russia was already conducting research on the uses of magnetism from a commercial sense. One of the areas being surveyed was the farming industry. Two discoveries were made that assisted with increased productivity –

1. Crops of lucerne that were watered with water which had travelled over lodestone (naturally occurring magnetic rock) yielded a forty per cent increase in growth.
2. Dairy cows that were fed the lucerne generated an increased milk supply.

The point is made that the water was not developed into a homoeopathic substance, neither was it a gem essence – it was a naturally occurring therapeutic medicine! The chemical composition of lodestone is mainly iron and oxygen. Could it be that iron and oxygen were energetically being transferred to the lucerne and as a consequence, to the dairy herd? In support of this concept Haas (1992, p. 200) states

> "….. some of our soil is iron deficient so the plants grown or the animals grazed there may contain relatively smaller amounts, though this is not yet a major concern." And the following "The primary function of iron in the body is the formation of haemoglobin. Iron is the central core of the haemoglobin molecule which is the essential oxygen-carrying component of the red blood cell …"

Staying Healthy With Nutrition is an American book on nutrition, but Russia could well have a shortage of iron in the soil. The fact that the lucerne responded, as did the dairy herds, leads one to think that soils in Russia could also be deficient.

Information relating to the therapeutic uses of energy medicine, be it in the form of Flower Essences, Gem Essences or the actual crystals, is either channelled information or programmed information. The person doing the channelling is the recipient of such information. Three different channellers may receive the same information; on the other hand, they may all tune in to different information. Whatever the information received by the channeller, it is valid because the intent of the person is 'in the mix'. This results in a blending of the energies – energy emanating from the crystal or the flower and the energy of the channeller. The same applies if a crystal is being programmed. There are specific processes involved but always the intent of the programmer takes precedence.

Melody (2008, p. 42) succinctly outlines this point:

> "Without conscious programming of a mineral that one desires to use for an elixir/stone oil, all of the energies are transmitted, and no specific direction for these energies is delineated; this method is conducive to activation of "spring tonics" and for general daily use. Most minerals can be used in the preparation of elixirs and stone-oil."

When dealing with crystals I believe that, as well, there is the very essence of the crystal itself which is based on its mineral content. It is this latter point which I will continue to pursue, as this book is written with a view to the reader having access to minerals that are no longer available or are becoming harder to acquire.

CHILDREN OF THE EARTH

One of the oldest peoples on earth, Australian Aborigines, have an inseparable link to their country. The land is seen as their mother and they are the children – the bond between mother and child is unbreakable. Developing that analogy further it is not difficult to conclude that Earth provides their sustenance. Water is the blood of Mother Earth and its rocks and minerals are the bones.

Regardless of skin colour we are all Children of the Earth. What lies buried in the Earth is also to be found within our bodies. Quantity and quality of course vary. Medical science has not yet made this discovery but as we move further away from orthodox medicine new discoveries will be made in this field and realisation will lead to the understanding that good health lies within us all. However, it needs to be understood that not all soils contain the minerals we need and secondly the pH of the soil is not always in balance, thus preventing the uptake of certain minerals. Humans, plants, animals and soil bear a resemblance, in that to function effectively all require a pH of around seven. pH stands for potential hydrogen and is a scientific measurement to determine the acidity or alkalinity of a substance. The scale ranges from one to fourteen; a reading of less than seven indicates acidity; more than seven indicates alkalinity. The further the reading moves away from seven in either direction is an indication of greater acidity or alkalinity.

The mineral content of a human has approximately the same ratio as is found in Mother Earth. This statement implies viewing Earth as a whole. An example would be the shortage of selenium in Australian soils being compensated by selenium-rich soil in Arizona. Is it a coincidence that:

- Australia has one of the highest rates of skin cancer in the world;
- Selenium is used in the treatment of cancer;
- Australian soil is deficient in selenium?

Mineral imbalance in the body leads to imbalance both from a physical and non-physical point of view. Naturopaths and other natural therapists are now using minerals almost exclusively in some instances, to treat physical and non-physical ailments. Interestingly silica is found in vast quantities in the earth and is widely used in treatment of the integumentary system (skin), the largest organ in the body. Silica is also used in treatment of hair and nails, bones, ligaments and joint stiffness.

Examples of mineral use at a non-physical level are potassium phosphate and magnesium phosphate. Potassium phosphate is used for poor memory, lack of concentration, depression, apathy and weepiness. Magnesium phosphate is used in treatment of stress, tension, headaches and colic. Administration of any mineral for therapeutic purposes implies it is not available at a dietary level.

As human beings our origins lie buried deep in the mists of time. A few enlightened souls may remember their birth but the majority of people have no recall of this event. Our forefathers knew and remembered our origins, with solemnity and ritual, when they laid one of their own to rest. "We therefore commit his body to the ground; earth to earth, ashes to ashes, dust to dust." (The Book of Common Prayer, p. 190)

CREATION

To view Creation from The Big Bang perspective of Science or the Dreamtime of Australian Aborigines offers diverging concepts which have ongoing ramifications.

Coming from the Big Bang viewpoint, planet Earth is described in a concise, systematic way. The substance of which Earth, and all who reside here, is made, is explainable in a tangible, non-spiritual manner. By digging into the earth, mining the oceans, initiating space travel, modifying plant life, or altering species, the religion of science has brought numerous benefits to humankind.

For those whose belief of Creation is linked to the Dreamtime process the Big Bang theory is simplistic. Somehow the implication of an ancestral Dreamtime brings a sense of responsibility and respect for all that has been created. (Was it not our Ancestors, aeons past, who were the Creators?)

The Big Bang theory has escaped this concept. In its quest for superiority, the religion of science has omitted respect and responsibility from the equation. Lawlor (1991, p. 19) refers to "the ladder mentality". The ladder symbolises climbing to the top – white males are on the highest rung while the rest – women, children, other races, plants and animals are designated to the rungs below.

Whichever way we choose to view Creation, in particular Planet Earth, the end result is the same – we have a planet with a diversity of life, a diversity of terrain, a diversity of climate, and a diversity of breath-taking beauty.

Heline (1977, p. 27) relates Creation to the number four.

> "Four represents the Principle of Creation manifesting as the four elements out of which all things are created. ... The four elements, ... are recognised upon this physical plane as fire, air, water and earth. On the higher or invisible planes these elements are recognized as spiritual forces".

Heline elaborates further

> "Four letters compose the Sacred Name of nearly all the Gods who have been worshipped by the human race. Note the following: Isis – Egyptian; ... Deus – Latin; Odin – Scandinavian; Dieu – French; Gott – German; Zeus – Greek; Atma – Hindu; and Jove – Roman."

Lawlor (1991, pp. 31-32) also links number four to Creation:

- There are four directions – North, South, East and West.
- Australian Aborigines have a universe divided by the colours of the four earth pigments – red ochre, yellow ochre, white pipe clay and black carbonic soils.
- Science also claims the four-fold system of elements - nitrogen, hydrogen, carbon and oxygen – the basis of all organic substance.
- The human body comprises four main tissues – bone, blood, nerve and layered tissue.
- Science recognises four force fields – gravity, electromagnetism and the two nuclear fields.

- There are four races of people – black, white, yellow and red.
- The Native Americans have four different colours of corn – yellow, red, white and black from which they relate the four directions and the four different races of people.
- Amongst the oldest indigenous tribes of Africa their dances involve repeating each step four times. The reason given "To be in this world is to be four in spirit and in body, like the four winds and the four legs of the animals."

And Arrien (1993, p. 7) states:

> "… my research has demonstrated that virtually all shamanic traditions draw on the power of four archetypes in order to live in harmony and balance with our environment and with our own inner nature: the Warrior, the Healer, the Visionary, and the Teacher. Because each archetype draws on the deepest mythic roots of humanity, we too can tap into their wisdom …."

WATER

Life-giving, thirst-quenching, a solvent, a carrier of minerals (as well as toxins), the blood of Mother Earth, and until comparatively recently a commodity which we have taken for granted. Water was a substance not taken for granted by Viktor Schauberger, (1997, pp. 6-8) an Austrian forester and naturalist, who in the nineteen twenties warned of an environmental crisis as the result of deforestation and the subsequent abuse and loss of water which he maintained was not merely hydrogen and oxygen but a living entity. Through close observation Schauberger determined that water, if it was to be unspoiled and "alive", must be allowed to follow its natural courses. In its natural state a water course bends and curves, its banks covered by bushes and trees. Through years of patient and thorough observation Schauberger concluded that water, if it were to retain its energy and vitality, needed to maintain a temperature of plus four degrees Celsius. Water likes to be transported in coolness and in darkness. Where else could mature water be born than in the coolness and darkness of the old forests?

It was of great concern to Schauberger (1997, pp. 6-8) that with the felling of the old forests little or no thought was given to the effect this would have on the supply of water. This concern was in the nineteen twenties and it would seem that in the second decade of the twenty-first century not much has changed. Water is the commodity, according to some, that will precipitate global conflict. Vast areas of rain forest are now being destroyed on a daily basis, destroying habitat of wild life and, as Schauberger would claim, contributing to the paucity of water around the planet.

If forests are the birth place for water and humans destroy this birth place, water becomes endangered, leading to an overall detrimental effect – no forests, no water, no food, no life.

More recently a Japanese researcher, Masaru Emoto, published *The Hidden Messages in Water* (2004a) followed by *The True Power of Water* (2005). Through the mediation of photography he established the living attributes of water. The photographs are of crystals which form when water is frozen. Over a period of five or more years Emoto photographed hundreds of such crystals and from these was able to identify what was "good" water and "bad" water.

Emoto (2005, p. 17) maintains that all water circulating on the planet does so in cycles varying from "thirty to fifty years". Because the last six decades have shown increased industrialisation and, as a result, increased air pollution, ground water now in use will remain polluted.

Skow and Walters (1991, pp. 16-17) explain with simple arithmetic the power of water as it relates to farming. The quantity of fertiliser used on one acre is comparable to some of the basic principles of homoeopathy namely: the Law of the Minimum Dose, or, Less is More. According to their calculations there are approximately five million atoms in one drop of water. Skow and Walters elaborate on these figures in relation to fertilisation, but none the less as far as water is concerned, it becomes exciting to contemplate the amount of energy that five million atoms generate in one drop of water!

Water has been used extensively for many years for energy medicine. Such modalities as Flower Essences, Homoeopathy and Gem Essences use water as a carrier of the energetic frequencies which emanate from the gross substances, for example, flowers, herbs, and minerals. Prior to these frequencies being introduced, the water is 'energetically' cleansed. Magnetic sheeting is placed under a glass container of water and left for several hours. Plastic or metal containers are unsuitable. I suspect this works much the same as placing a credit card next to a magnet –

the magnet wipes off the signature and information. I have been using this procedure for many years not understanding, or even being aware of, the wider implications of magnetised water.

Schauberger (2000, p. 2) states that through over-illumination by sunlight and consequent over-warming of the ground, plus various methods of water storage which are not natural, water has become demagnetised and "The demagnetisation of water signifies the removal of its soul."

Martlew (1994, Overview xii) states

> " …. drinking water, once our most important source of minerals, is now an insignificant contributor to our daily intake because it no longer comes directly from lakes or streams; by the time it reaches us it is almost as processed as our foods. Water running over rocks picks up minerals, and the swirling of the water creates vortexes which 'charge' the minerals it carries. The highly processed water running through pipes and into our faucets hasn't got a spark of electrical energy left in it."

When one thinks of how water is stored – mainly in dams – Schauberger's comments regarding water's temperature come into stark focus. Australia is the driest continent on Earth and now in the twenty-first century we are faced with the dilemma of having insufficient water, our rivers are either drying up or have become toxic, and our rainfall is erratic. It is difficult to find the 'water running over rocks, picking up minerals and the swirling of the water creating vortexes which charge the minerals it is carrying.' In various parts of Australia we are building desalination plants to service the increasing demand for water.

If the overall supply of 'living' water on the planet is being depleted the implication is that food also will be in short supply, as will minerals. I refer to the point made earlier – is it possible to obtain minerals directly from the rocks or stones by firstly cleansing or magnetising water then submerging a specific rock or stone in that water to extract the frequency of the mineral(s) contained therein?

Prior to the toxicity of waterways, when the planet supported pristine forests and the water was rich in minerals and energy, the health of its people was not in contention. Food consumed came from mineral rich, healthy soil, which was watered with life-giving and mineral rich water.

Over the past two or three hundred years, due to population increase, farming methods have changed to accommodate greater food production. Water, likewise, has undergone various changes. Purification has taken the form of chemical additives while storage methods, due to population increase and demand, have diminished the purity of water. Society may not be prone to as many water-borne diseases but the essential quality and purity of water has lessened. Added to this situation environmental pollution has become an increasing problem.

I maintain 'good water' that Emoto writes of (2004a, 2004b & 2005) is therefore essential to human health. Emoto's research and photographs of water crystals give countless examples of this phenomenon. After many years of research and extensive photography, Emoto has established that water has a memory whereby information is stored and transferred. He cites as an example (2004a, p. 88) the first Gulf war. Measurements taken of vibrations of water in Tokyo on the afternoon of the Iraq invasion revealed extraordinary increases in the measurement of substance vibrations, such as lead, mercury, aluminium and other toxic materials. Repeated tests revealed the same results and it was not until the next day that newspaper headlines revealed that the invasion of Iraq had begun. He states that the number of bombs dropped during the first day of the invasion equalled that of the entire Vietnam War. By-products of the

Iraq invasion did not reach Tokyo; however, the vibrations of harmful substances did, and the water in Tokyo absorbed those vibrations.

In the light of Emoto's research, the ability of water to store information prompts the question – Is there any accessible 'good water'? This planet has experienced countless wars which have involved toxic substances on a massive scale – has water around the planet absorbed, at an energetic level, the vibrations of those substances?

In the late nineteen nineties research was conducted by a group (Kronberger et al, 1995, p. 6) in Milan on water obtained from several places where the water was regarded as miracle water. Fatima, Lourdes and Medjugorje were among several centres selected. Each of the samples taken from these centres recorded seven frequencies. Prior to these tests the group conducted tests on tap water in Milan and an alumina deposit adjacent to the miracle water in Medjugorje. Results of both the tap water and alumina deposit tests indicated that no frequencies could be measured. After the miracle water tests, the tap water and the alumina tests were measured again and found to contain those same seven frequencies as in the water from Medjugorje. This gave rise to the interesting possibility concerning transfer of information. Remnants of water had been stored in equipment which was used later in the re-tests.

During the latter half of the twentieth century Johann Grander, an Austrian naturalist, conducted considerable research into the energetic properties of water. Grander's work and that of Viktor Schauberger, it could be said, were the way-showers for researchers such as Emoto, Ciccolo (from the Milan group), and Beneviste the French biologist. Grander remained close to and was a keen observer of nature throughout his life. His constant devoted interest in nature, particularly water, led to his development of revitalisation of water. Through magnetism water is subjected to high frequency vibrations and according to Grander (in Kronberger et al, 1995, pp. 57-58) these vibrations are information. He elaborates further on the concept that every living thing, including minerals, resonates with the cosmos and thus also with its own planets. Planetary energy is drawn, changed, refined and the excess discarded. Water, according to Grander, is the significant bearer of information for these vibrations. We can conclude from this the significance of water and just how closely everything on earth is linked to water. Grander explains how genetic information is within every seed and cell as it is in water itself. Seeds may be stored for decades, even longer, without sprouting. Once water is added the seed sprouts and growth commences. Grander claims that the information contained in the water activates the elemental information within the seed and not, as most people think, the moisture that activates growth.

Crop planting and subsequent seeding poses interesting questions in the light of Grander's findings. Does rain containing a myriad of toxic vibrations, influence in any way the germination of seed? Is this one reason why crops require increasing amounts of fertiliser to flourish? Does it depend where on the planet the crops are grown? Is this what is known as acid rain? Or does acid rain actually contain the gross toxic substances?

Mention was made earlier of the use of magnets for cleansing water. I use water which has been cleansed by magnetic sheeting for making homoeopathic substances – the idea being to remove unwanted vibrations from the water before imprinting the vibration of the homeopathic substance I require. What occurs here is similar to the water Emoto creates; he super-imposes his message of say 'love' and 'gratitude' which eliminates the adverse or low frequency vibrations from the water. The piece of magnetic sheeting does the same; it leaves the water free of low frequency vibrations and the vibration of the homoeopathic substance takes precedence.

To elaborate further on this concept, instead of imprinting the water with a homoeopathic substance, placing a crystal into water which has already been cleansed of adverse frequencies by the magnet, would infuse the water with the frequencies of the crystal. This in itself is not new. What is cause for excitement is that 'clean' water is being used for the crystal infusion – the crystal frequencies are not in competition with other frequencies which may have been held in the water.

During the writing of this book I adopted the same principle. The water I drink is rain water which has been filtered to remove any toxic material substances. In addition I placed a piece of magnetic sheeting underneath the jug of 'clean' water; the magnetic sheeting stays beneath the water jug all the time. I then pour from the jug into a drinking glass in which I place a crystal that I have selected to work with. To date I have used separately lapis lazuli, nuummit and sodalite. I have no doubt these three crystals have assisted my thought process whilst writing.

If a hexagonal crystal was to be used in the same way, then surely this would offer maximum benefit to the human body bearing in mind that our body is more than seventy per cent water and the statement made that our DNA molecule is hexagonal (Narby, 1999, p. 130). Surely a drink of such water would provide nectar to our cells.

MAGNETISM

Mother Earth herself is a gigantic magnet with a North Pole and a South Pole. Circulating between the two poles is the energy of the magnetic field which the two poles have created. According to studies carried out by Davis and Rawls (in Holzapfel, Crépon & Philippe, 1986, p. 22) the energy generated by the two poles differs. Japanese scientists claim that the intensity of the magnetic field is decreasing and have estimated that terrestrial magnetism has diminished by fifty per cent over the last five hundred years. Other scientists claim that although the Earth's magnetic field may have weakened, it could regain strength in coming centuries. Some scientists have put forth the concept that the present weakness of the Earth's magnetic field is responsible for the many illnesses and that this deficiency is harmful to all life.

Schauberger's (2000, p. 2) comment that water has become demagnetised is not an isolated observation. Martlew (1994, p. xii) claims that because of the way in which water is stored and processed it no longer holds any electrical energy.

Further research during the last thirty years by people from various parts of the globe has offered more precise data on the magnetic properties of water. Russia's research pursued an industrial path while India, France and England probed the possible effects on human health. Exciting results emerged to the extent that today the use of magnetised water is widely practised.

Schauberger's observation more than seventy to eighty years ago begs the question about humanity's health at that time. If the quality of the water was deteriorating then and subsequently the widespread use of chemical fertilisers became the norm, is it not likely that there would be an acceleration in the downward spiral of human health?

There is an analogy here with the health of the water (the blood of the Earth) and the health of blood in the human body. Studies by Drs H L and R S Bansal (1985, p. 37) accentuate this point by their findings that the quality of a subject's blood was significantly improved by magnetic flux created in the blood.

FARMING PRACTICES AND SOIL

To say humanity has benefited because of innovative technology and sophisticated farming methods is a superficial comment. Yes, there have been enormous advantages, but at a deeper level there has been impairment and in some instances downright harm which has been brought about by impoverished soil and subsequent mineral deficiencies in both livestock and people. Farming methods over the last hundred years have put a new face to agriculture, quantity has replaced quality, bigger has to be better. Because of this, white society in particular has moved away from or even lost the art of living in synchronicity with the seasons.

To keep up with demand for food, techniques of supply had to be improved. Bigger and better farm machinery was brought into use, larger tracts of land were cleared and that magical substance superphosphate was put to good use. Initially the yield of grain per acre (or hectare) exceeded all expectations but after a period of around four to six years the amount of grain harvested reduced quite considerably. The land was not responding so farmers cleared more, results were similar – for the first few years crops were exceptional then the decline began once again. More land clearing meant more deforestation which led to more soil erosion. According to some authorities on water (Schauberger, 1997, pp. 2, 8) (Suzuki with McConnell, 1997, p. 102) continuation of this sequence has led in part to water shortage.

As a consequence of overuse of superphosphate soil became acidic. Once this happened nutrients in the soil became unavailable to plants and therefore to humans. Put simplistically, humans, animals, plants and soil bear a resemblance in that to function effectively pH needs to be around seven. Movement to either side away from seven results in an acidic or alkaline environment, thus inhibiting the uptake of nutrients.

Soil requires an environment conducive to supporting beneficial micro-organisms. Main constituents are oxygen, food, water and comfort. If such an environment is not available these beneficial micro-organisms are unable to survive resulting in cessation of life – including human.

ROCKS

Rocks are the bones of the earth. Like humans with their skeletal framework rocks provide material to which, in some form, everything is attached, be it mountains, valleys, ocean beds, or rolling hills. Beneath this vast landscape/seascape lie the tectonic plates which it is generally accepted, are responsible for dramatic movement such as volcanoes and earthquakes. Tectonic plates are constantly on the move and in terms of human movement the rate is miniscule – something like one centimetre or less than half an inch per year. Twenty times faster than this is the Nazca plate in the Pacific Ocean with movement of something nearer to eight centimetres per year. What causes plates to move is still being determined – the most common theory being the concept of convection in the Earth's mantle.

Earthquakes occur where plates shudder past each other. Volcanoes occur mainly close to plate margins and where there is a good supply of magma which is created by the melting of mantle rock. This then bubbles or wells up to the surface and erupts as a volcano. Collision of plates causes edges of continents to crumple throwing up mountain ranges. Plates themselves are gigantic slabs of rock, the largest of which is the Pacific Plate underlying most of the Pacific Ocean. It is not difficult to imagine that planet Earth is never still. Relentless movement of the plates, coupled with endless weather cycles, earthquakes, tsunamis and volcanoes bring the realisation that we live on a planet undergoing continuous change.

From this chaos of never-ending movement emerge the bones of the Earth – Rocks – of which there are four types:

- Igneous
- Sedimentary
- Metamorphic
- Conglomerate

Geologists refer to a Rock Cycle, a continuous process which involves breaking down and re-creating rocks. Exposure to weather – frost, sun, wind, rain and sudden exposure to extremes of temperature – contribute to erosion. Of greatest significance is water, providing constant movement, wearing away material from one place and depositing it elsewhere. Rivers provide an example of how, over millions of years, canyons are formed. Arizona's Grand Canyon and gorges in North Western Australia are illustrations. Waves constantly crashing on the seashore act to carve out sometimes dramatic shapes of rock. Whenever rock surfaces are exposed action of erosion begins, thus new rocks evolve and develop. A gradual unstoppable re-cycling process of mammoth proportions commences, maintaining its purpose over millions of years.

New rocks are also created when eroded material sinks back down into the Earth's mantle. While it may seem this material is lost, such is not the case. The re-cycling process continues and millions of years later it re-emerges as molten magma. (Magma is molten rock below the surface of the Earth while lava is the molten rock which is ejected at the surface of the Earth.)

Farndon (2007) explains the different types of rock as follows:

Igneous rocks are subjected to more change than any other rocks and form a large section of the ocean crust and continental crust. Igneous rocks mainly consist of hot molten rock which has been frozen solid. The freezing point of magma is much higher than that required to turn water into ice. The temperature required for water into ice is zero degrees Celsius (thirty-two degrees Fahrenheit) while magma freezes at temperatures between six hundred and fifty degrees Celsius (twelve hundred degrees Fahrenheit) and eleven hundred degrees Celsius (two thousand degrees Fahrenheit). The process of freezing or solidifying is not something that occurs uniformly as magma contains a number of substances such as magnesium, iron, silicon, and potassium, each with its own freezing point. It is at this point, when the magma is cooling, that elements and compounds link to form crystals.

Sedimentary rocks are easier to distinguish from igneous and metamorphic rocks because they form in layers. There are two features that differentiate sedimentary from metamorphic and igneous rocks. The first is that a sedimentary specimen will usually break along its layered surfaces, and secondly these rocks house a variety of fossils. Rarely are fossils found in igneous and metamorphic rocks. Fossil content, be it plant or animal, provides a key to the rock's origin and to its age.
Sediment lying on the ocean floor or beds of lakes and rivers provide the beginning of the process out of which sedimentary rock is formed. From pressure caused by build-up of layer upon layer of sediment over millions of years plus intense heat from Earth's interior we have the creation of sedimentary rock.

Metamorphic rocks, as the name implies, have evolved through change. Geologists realised that igneous rocks were the result of melts and sedimentary rocks developed from sediment. They have now realised that any rock – igneous, sedimentary or metamorphic – can be changed into new rock. If a rock is subjected to extremes of pressure and heat it would alter, but not so that it would melt or break down completely. Should the heat have been intense enough to melt the rock an igneous rock would have been created.
Geologists have studied metamorphic process and an example of this change was found in marble, which appeared to be part igneous and part sedimentary rock. Heat altered the original limestone into calcite and its crystals when reformed resembled igneous rock. Pressure and temperature determine the type of metamorphic rock. Examples of this are: low pressure results in slate; high pressure and high temperature with circulation of fluids result in gneiss; heat alone creates sandstone.

Conglomerates. As the name implies, these rocks/stones contain a mixture of stones. Usually rounded stones such as pebbles, cobbles, even boulders are set into rock, sometimes resembling a cake mixture. In geological terms this "cake mixture" is referred to as a matrix. Conglomerates are mainly formed from sedimentary rock and result from activity or movement associated with swiftly flowing rivers, landslides, glaciers and avalanches. To survive the constant buffeting, stones in a conglomerate are usually of a quartz or flint type.

THE CELL

With the advent of the electron microscope cell biologists are now able to shed light on the behaviour of cells at a molecular level. What the "new age" folk have been insistent and unswerving about for decades, and indigenous cultures for centuries prior, cell biologists are now offering explanations from a scientific viewpoint. It appears that science will soon be able to confirm what intuition has known for millennia. This can be regarded only as a positive step forward; it will assist in the breakdown of the "them" and "us" mentality.

Meaningful research on crystals and their medicinal uses is now within reach of those who desire to extend their boundaries of healing to encompass this modality at a traditional as well as an allopathic perspective. Trying to establish a connection between crystals and human beings has been, and still is being, researched. The findings are interesting and support the popular metaphysical belief that human beings and crystals do have a connection and a strong one at that. Some biophysicists have put forth the concept that human beings and other organisms are liquid crystals. One definition of a liquid crystal is that the liquid contains within it component particles or molecules that have been arranged in such a way that their orderliness surpasses that of ordinary liquids and comes close to solid crystals. There is an unbroken mass of connective tissue in the liquid crystalline cytoplasm which forms part of the interior of every single cell.

On that basis it becomes clear how the human body operates as a cohesive coordinated whole. Through this liquid crystalline substance that permeates the whole body absolute and total communication is achieved from organs down to the interior of every cell.

Another discovery which adds strength to the idea of a strong connection between crystals and the human being is the claim (Narby, 1999 p. 130) that the DNA molecule existing in each cell has the structure of a hexagonal crystal (or is a hexagonal crystal). These findings provide strength to my theory that crystals provide an optimal means of mineral energy transference to the human body.

"At a cellular level" – what a mysterious yet revealing place to be! This is where life begins, this is where things happen, this is where the natural and supernatural inter-fuse, where physics and metaphysics blend. It happens within every cell of the human body and there are trillions of them.

As a result of courageous researchers like Bruce Lipton, MaAnna Stephenson and Joyce Hawkes it will be possible for complementary and alternative medicine practitioners and others to be acknowledged and recognised by main stream health professionals, enabling them to pursue their chosen path without the denigration which until now has accompanied them on their journey. As well as the scientific advancement, such research will lead to

1. Acknowledgement of the wisdom which indigenous peoples have worked with for millennia;
2. Recognition of the part which minerals and water have in the nourishment at all levels of the human body;
3. Enhancement of skills from both sides of the "great divide" leading to an increased capacity to heal, thus benefiting those who are in need of healing; and
4. Greater acceptance and understanding between practitioners of all persuasions.

The Human Cell

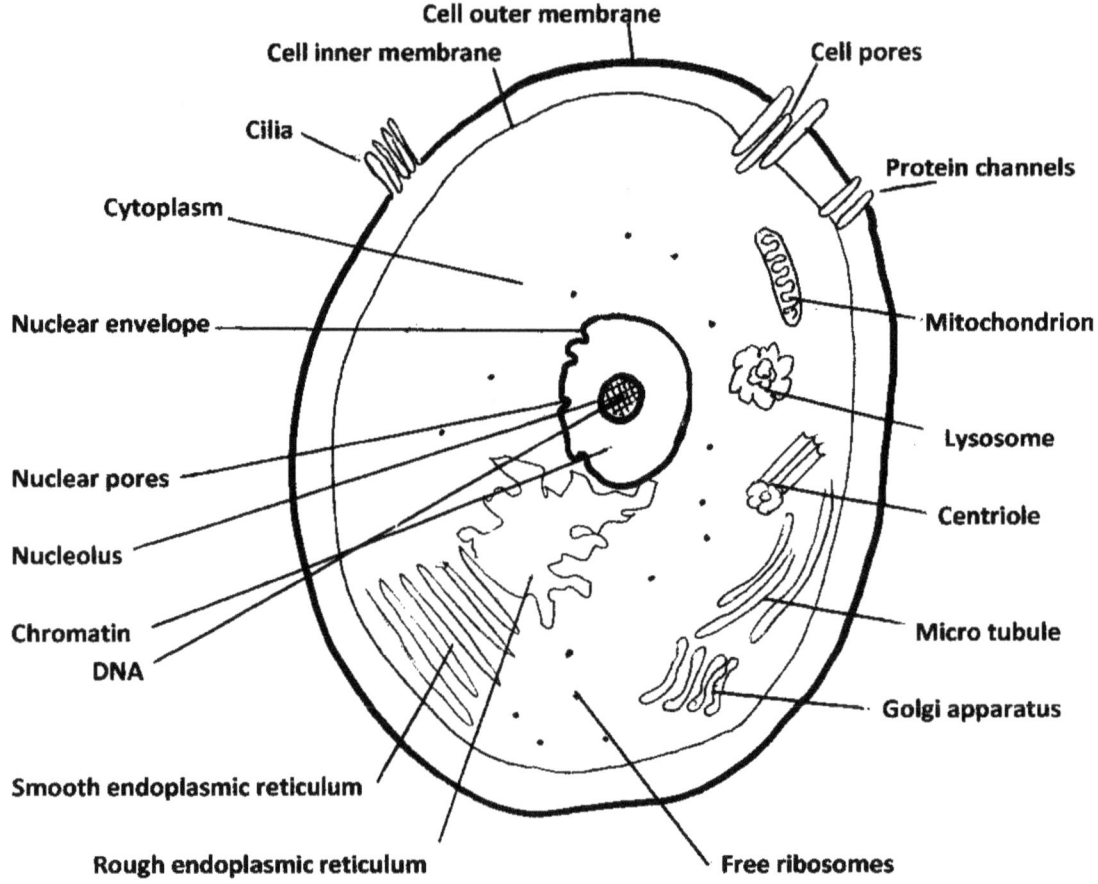

Lipton (2005, p.39) describes the human body as an enormous community of cells, each cell being autonomous but never the less functioning altogether as a whole. This concept lends itself to the idea of both the simplicity and complexity of the human body. Life started as a single celled organism and there is evidence to show that these single celled organisms were here on Earth within six hundred million years of the Earth being formed. For approximately two and a half billion years these single celled organisms occupied the Earth. Around seven hundred and fifty million years ago single cells linked to form a multi-cellular entity. This was a turning point as cellular communities then organised themselves expediently into larger and larger communities, some consisting of millions, some of billions and even trillions of single cells. While a human being, a cat, dog or elephant appears to us as a single entity, the truth is these creatures are a cohesive assemblage of cellular communities.

Development of the electron microscope has made it possible to examine the single cell, enabling researchers and scientists to expand their understanding of life and in so doing offer hope in a world where survival for those in poor health is increasingly difficult.

Within each cell are many components known as organelles or miniature organs, each performing set tasks which enable the cell to function efficiently. Hawkes (2007, pp. 90-91) states "There are five components of cells that need to be understood in order to activate your intrinsic capacity for renewal and healing." Hawkes then lists these five components and their functions:

Cell Membrane – the cell boundary and the receptor sites which assist with communication and aid the regulation of the cell's activity in relation to the passage of materials both in and out of the cell.

Nanotubes – microscopic extensions of the cellular membrane that connect to neighbouring cells. Their task is one of communication in that they transfer information between cells.

Nucleus – the membrane-enclosed kernel within the cell which houses the DNA which holds the job description for all cells within the human body. Sometimes referred to as the Control Centre, it houses the genetic material and cellular activity including cell reproduction.

Endoplasmic Reticulum – consists of a network of specialised membranes in the cytoplasm (the watery substance within the cell that holds everything together). The task of the Endoplasmic Reticulum is the final production of specific proteins.

Mitochondria – are the microscopic power houses within the cell. Through a system which involves its own DNA and specific enzymes these power houses produce the energy needs of the cell.

Lipton (2005, p. 37) puts forward the concept that mechanisms "employed by cellular organelle systems are essentially the same mechanisms employed by our human organ systems ... there is not one "new" function in our bodies that is not already expressed in the single cell." Lipton elaborates further that each nucleus containing cell has the capacity to operate in ways comparable to our digestive, respiratory, endocrine, muscular and skeletal, reproductive, circulatory, nervous and integumentary systems.

The research offered here provides an understandable yet simple explanation of how humans function. Our body consists of trillions of cells arranged in an orderly fashion. Different cells perform different tasks and these are arranged accordingly, for example cells of the respiratory system are different from nervous system cells. Thus each system is its own well organised functioning community.

Emoto (2004b, p. 110-111) offers a striking analogy when he refers to the human body as a megalopolis, the city of the body. He compares the world population of approximately six billion to the human body with its approximately eighty billion cells, each with cells in constant communication with each other. This huge city runs on a transport system which is controlled by water. Most of the city's requirements are brought in by water - a river running through the city transports the necessary cargo which is dispensed throughout the entire megalopolis. This system of distribution ensures efficiency and effectiveness. Sanitation is taken care of with the same degree of organisation. Barges further downstream, loaded with waste material, leave the city on the outgoing tide. In conclusion however, Emoto asks why there are so many human beings in poor health if the megalopolis – the city of the body – is so efficient. He suggests that the lack and poor quality of the water in the city are responsible for a break-down in the transport system.

The work of Emoto has added another dimension to the graphic image outlined by Lipton. Lipton's research has provided an accessible and understandable pathway to knowledge of human cells.

His simple but brilliantly informative explanation offers the reader the opportunity to actually co-create with their cells. No longer is knowledge of the human cell exclusively the domain of science; Lipton invites the reader to explore the stuff of which humans are made and engage in a collaborative relationship at a cellular level. Add Emoto's imaginative portrayal of a vast city powered by water and we are left with the impression that each of us is in control of our own city. As the Director of Operations of this vast body of cells (your own city) you can communicate with each and every one of them, you can control the flow of the river, you can determine the quality of water in the river, you can determine the state of health within your own city.

CRYSTALS

The energy/vibration emanating from the crystal is dependent on its size, the particular shape of the crystal, its colour and its mineral content.

There is little doubt that a particular crystal, when one uses it for metaphysical purposes, assists in relieving depressive states, anxiety, fear, doubt, jealousy.

Is this because of its shape? Its colour? Its size? Its mineral content? Or its programming?

Could it be that these metaphysical properties emanate from its colour, its shape or could it be that it is in fact, because of the mineral content? Take copper as an example – the blue-green colour in certain crystals is due to its copper content. In the human body (both physical and non-physical) copper (in excess or deficiency) is linked to certain conditions such as heart ailments, arthritis, premenstrual syndrome (fatigue, depression, emotional instability). Examples of copper containing minerals are bornite, chrysocolla, covellite, and cuprite.

CRYSTAL SYSTEMS

Crystal Symmetry is the language used to describe the three-dimensional aspect of a crystal. These are the Axes, Planes and Centre of Symmetry. Gemmologists and lapidaries require in-depth knowledge of crystal symmetry to enable greater insight into the cutting and visual properties of crystals.

There are seven systems: Trigonal, Cubic, Hexagonal, Tetragonal, Orthorhombic, Monoclinic and Triclinic. Energetic features of each system differ. Gerber (1988, pp. 353-356) explains the cubic system, to which diamonds, members of the garnet family and fluorite family belong, as having energetic properties which lend themselves to the repair of bones and cellular structure. Crystals of the tetragonal system, to which chalcopyrite, diaboleite and cassiterite belong, display energetic properties which act to balance the positive and negative energies. Stones of the triclinic system, such as chalcanthite, turquoise, pectolite and rhodonite, have energetic properties which act to balance the yin and yang energy. The trigonal system includes crystals such as dioptase, smithsonite and benitoite. Its energetic properties act in a spinning motion, they are neither positive nor negative. The monoclinic system to which lazulite, vivianite, mimetite and malachite belong, display an energy which expands and contracts much like pulsation. The orthorhombic system includes crystals such as chrysocolla, topaz and celestine. Energetic patterns involving thought forms, projection and protection are synonymous with this system. Hexagonal crystals such as members of the beryl family, for example emerald, aquamarine, plus other crystals such as apatite and zincite, present energies which boost growth and vitality. Similar to the energies of the cubic system, hexagonal crystals also possess an energy which is beneficial to the physical human body as well as energies of the meridian and chakra systems.

The precision and accuracy of crystal symmetry leave little doubt as to the absolute perfection of the universal creation. Within the cubic system there are nine planes of symmetry; the hexagonal system contains seven planes; the tetragonal system five planes; the orthorhombic system three planes; the trigonal system three planes; the monoclinic system one plane; and the triclinic system has no plane of symmetry.

HEXAGONAL CRYSTALS AND SHAPES WITHIN NATURE

Numerous hexagonal crystals are discussed in this book, however, master amongst them must be the clear quartz crystal. Considerable information has been recorded regarding this crystal and knowledge of its uses go back into past civilisations such as Atlantis, Lemuria and Mu. Anthropological writings reveal wide spread use of quartz amongst the indigenous cultures of North and South America, India and Australian Aborigines. A common thread running through all the literature reveals the power of the quartz crystal both as a healing stone and an amplifier of energy. The fact that quartz crystal is found worldwide leads one to think that its uses for humanity have not yet been fully discovered. The energy of this stone is remarkable in that it can store, transmit, receive, focus and regulate energy as well as become attuned to the unique energy of the individual requiring healing. Clear quartz works on all levels: physical, mental, emotional and spiritual. It acts as an immune stimulant; assists with concentration; may be used as an effective receptor for programming; and is used in the development of psychic capabilities. Kirlian photography has shown that a person's bio magnetic field is doubled when a quartz crystal is held in the hand (Hall, 2003, p. 225). Melody (2008, p. 603) makes an interesting statement about the use of quartz crystal as a means of communication with other kingdoms such as plants, animals and minerals. Melody further extends that concept explaining that in past times when everything was thought to be connected, the quartz crystal was used to synchronise the individual consciousness with the total consciousness and the heavens.

Nature is an economist where competence and energy conservation are concerned. There is no place in the scheme of things for waste and inefficiency. Hexagonal shapes within nature are not uncommon and highlight the absolute proficiency and effectiveness of this six-sided shape. Author and educator, Schneider (1995, p. 179) refers to structure-function-order when describing hexagonal phenomena in nature. Examples Schneider (1995, p. 192) offers are star coral, tube coral, scales on fish and skin of reptiles. The hexagonal shape offers maximum strength with a minimum of energy.

Bees are the most famous for their use of the hexagonal style of building. Their honeycomb is constructed in alternating hexagonal tiers. Wax is produced by the bees after they have consumed large quantities of sweet substance. It is then chewed until it is soft and workable then placed in the hexagonal shape within the hive. The whole process is economical and efficient with shape, energy required and amount of material needed for the task; a little under two kilograms of honey is stored within forty-five grams of wax (Schneider, 1995, p. 196).

Within the human body, striped muscle, which is required for voluntary movement, consists of cells packed into a hexagonal pattern. Anthropologist Narby (1999, p. 130) refers to the four DNA bases as being hexagonal.

Emoto's illustrations (2004a, 2004b, 2005) demonstrate the hexagonal crystals which occur in 'good' water.

Bourgault (1997, p. 40) compares the hexagonal structure of quartz crystal with the hexagonal structure of the energy field surrounding the human body. He comments further that quartz crystal (SiO_2) is one of the most commonly occurring crystals on the planet and silica comprises approximately sixty-five per cent of the Earth's composition.

References to crystals (not necessarily hexagonal) within the human body are not uncommon. Cell biologist Lipton (2005, p. 90) refers to two types of crystal: hard and resilient, like diamonds; and liquid crystals, whose molecules form an organised pattern. Lipton refers to the cellular membrane as a liquid crystal. Oschman (2000, p. 129) states that crystalline arrangements within living systems are not the exception but rather the rule. Oschman cites connective tissue, myelin sheaths and phospholipid molecules forming cell membranes as examples.

Stephenson (2008, p. 44) explains that the organisation of DNA in chromosomes has a liquid crystal nature. M P Hall, (1972, p. 217) when discussing the pineal gland, refers to small granules or crystals contained within that gland.

HYPOTHESIS

My hypothesis is therefore that if the crystalline substance within cell DNA is hexagonal the use of hexagonal crystals, for either physical or non-physical purposes, in combination with 'good' water (hexagonal crystals), offers enhancement to the cells, and therefore can only be of benefit. The 'good' water increases sympathetic resonance which provides greater amplitude.

The following support my hypothesis:

1. Within nature there is an abundance of the structure–function–order of the hexagonal shape. Numerous examples of this have been noted earlier.
2. Masaru Emoto's extensive photographs of hexagonal crystals contained only in the good quality water offer further confirmation. Reference is made to this in the section on water.
3. The concept that the DNA within the human body is itself a hexagonal crystal (Narby, 1999, p. 130).
4. Use of hexagonal crystals by indigenous peoples of the world from the Amazon Basin to the Australian Aborigine.
5. Snowflakes – all different – are hexagonal.

Supporting concepts three and four, Narby refers to the four DNA bases as being hexagonal but each varies slightly in shape. Other descriptions of the DNA double helix offered by Narby refer to two serpents entwined. In his search for connection between the cosmic serpent and DNA Narby establishes a link between the two: the cosmic serpent lives in water and is both long and small, single and double and the DNA fits that description (Narby, 1999, p. 86).

NUMEROLOGY AND CRYSTALS

The universe is mathematically designed and is mathematically precise and as a consequence of that there is a place for everything within the whole. Ancient and modern philosophers have not discounted this possibility.

Pythagoras, born in 582 BCE and died 507 BCE, was the first man to call himself a philosopher and according to Lewis (1984, p. 32)

> "Number theory was fundamental to the teachings of the Pythagorean Brotherhood at Crotona. According to their ideas, number is the essence of the created universe. Number is Being. The cosmos was created and ordered according to the divine, ideal plan or pattern. Number is basic to the nature of the divine pattern and its manifestation

in the actual world. Because it is the basis of creation, it is also the fundamental nature of the law of correspondences. Number, creation, cosmology and music are all related."

Suzuki with McConnell (1997, p. 12) make a succinct statement when referring to a worldview in which everything within the universe connects with everything else. Forests, oceans, stars, humans, clouds - all interconnect; nothing does or can exist in isolation.

If one can accept the forgoing theories then it should not be difficult to accept that there is nothing haphazard about the world of crystals. The cubic system of crystals is an outstanding example of precision and perfection as it has nine planes of symmetry.

Heline (1977, pp. 74, 76) introduces the number nine as follows:

> "Nine is the emblem of matter which, while changing and in constant flux, yet retains its identity and resists complete destruction. This is manifest in the strange phenomenon of 9 remaining 9 in its power no matter by what number it is multiplied. It eternally reproduces itself." and "… revealed by the numerical truth that all numbers from 1 to 9 reduce to 9. 1 plus 8 equals 9, 2 plus 7 equals 9, 3 plus 6 equals 9, 4 plus 5 equals 9. Thus 9 is truly the number of matter, the number of man's evolution and the number of cosmic knowing …" and again "The numbers 1-2-3-4-5-6-7-8-9-10 equal 9. Also 9-18-27-36-45-54-63-72-81-90 equal 9. Nine when multiplied by another number always reproduces itself; … It indicates at once its power and its universality."

From this explanation of the number nine it is, perhaps, explainable why one of the crystal systems should feature nine so prominently.

According to Gerber, (1988, p. 353) nine's energy is reflected by cubic system crystals which have "an energy pattern which can assist in repair of cellular structures which have been damaged. This can be from the molecular level of DNA through to the bones of the skeletal system. The energy tends to be of an earth-plane nature."

If the number nine is as powerful as Heline suggests it would seem that this is in line with what the cubic energy system is capable of achieving. Here we are looking at gross physical matter – "bones of the skeletal system" and the cubic crystals' energetic abilities to change matter.

Looking at another example, namely the orthorhombic system, which has three planes of symmetry, Heline (1977, p. 18) has this to say of three:

> "The force and power of three has been identified with the Trinity by the Wise Men throughout the ages. All great world religions worship a three-fold Godhead. It is one of religion's fundamental teachings. In Christian terminology the three persons of the Godhead are the Father, Son and Holy Ghost."

On a less religious note, three is considered to be the number of perfection.

Another way of explaining the power of three is: (Heline, 1977, p. 18) "One projects from itself 2; from the component parts of 1 and 2, 3 is formed."

Looking at the energy of the orthorhombic system Gerber (1988, p. 354) points out that this system has a unique way of being able to encircle and encompass energy patterns and thought forms. Here we see that the energy is capable of organising the non-physical (energy patterns, problems and thought forms) and in so doing perhaps exerts a measure of control.

From these two examples we may understand something of the power of numbers. This power is not only a quantative power, but a qualitative power also.

If the reader is interested in further pursuing the science of numbers I recommend *Behind Numerology* by Shirley Blackwell Lawrence, published by Newcastle Publishing Co Inc, North Hollywood, California, 1989.

THE DINSHAH CONNECTION

This work would be incomplete without reference to Dinshah P Ghadiali (1873-1966). Born in India of orthodox Parsee Zoroastrian parents Dinshah commenced his schooling at two and a half years of age. His insatiable thirst for knowledge led him, at the age of seventeen, to become a scientific lecturer, having already been appointed as a research assistant at age eleven to his professor of mathematics. Dinshah's studies included chemistry, electricity, sound, heat, light, magnetism, hypnotism, mesmerism, engineering, medicine and watchmaking. At the age of thirty-eight Dinshah arrived in the United States, from where, nine years later, he launched his method of colour healing. It was from around nineteen twenty-five that the American Medical Association and the Food and Drug Administration began to show more than a passing interest in Dinshah's work. From then until nineteen fifty-eight Dinshah was the subject of numerous court cases, imprisonment, and continual harassment. He was prevented from selling his books or projectors and was confined to New Jersey where he continued to study and research. He died in nineteen sixty-six at the age of ninety-two.

The essence of Dinshah's work was the fundamental connection between the elements, and consequently minerals, and colour. Contrary to scientific theory Dinshah's research led him to conclude that chemical elements were in fact compounds. He demonstrated with hydrogen which emits two colours; the first being pale blue followed by bright red, which is the distinguishing colour of that element. Dinshah was working in the first instance with the dominant colour rather than the elements themselves per se.

Dinshah's ideas (MacIvor and LaForest, 1979, p. 62) took him further on the colour/element connection. He postulated that if the human body contained the necessary minerals for survival then it was logical that the human body would also contain colour bands or waves. His theory was that an imbalance of colour within the body implied a mineral imbalance. This imbalance would be rectified by infusing the body with a specific colour which would introduce the corresponding element or mineral into the body. Restoration of balance at the energetic level ensured that physical balance would be restored.

In spite of the ongoing feud with authorities in the USA, Dinshah had many supporters and practitioners who used his method of treatment. It is still a modality which is in use today.

Dinshah stressed that his system of colour healing does not deal with pigment colours or principles which may govern their combination but the use of radiant energy. The cornerstone of Dinshah's research is the relationship between colour and the elements and this is of interest to me in understanding the role of colour in relationship to crystals.

EMPIRICAL AND RECORDED KNOWLEDGE OF CRYSTAL USES

Reference is made concerning treatment with crystals where the crystal is placed underneath the tongue – a crucial reflex point. Gurudas (1986, p.15) explains that when a crystal is placed under the tongue, meridians and the midbrain, in particular the pineal gland, are activated. Bourgault (1997, p. 41) refers to the American Indians' use of amethyst crystal placed under the tongue for the treatment of addictions. Bourgault further explains how glands under the tongue are activated by the crystal and information is passed to the entire biochemical system within the body. (Placing a crystal under the tongue in this way is obviously not an option for children, the elderly, or intellectually challenged.)

These two references provide a brief insight as to the use of crystals in a healing capacity. Recorded history outlines the uses of crystals allowing us a glimpse of their medicinal purpose in times gone by. No doubt much information has been lost either in written or oral transfer of knowledge and we are left wondering "is that all?" I personally do not believe "that is all". I believe we are poised on the brink of re-discovery which in turn will precipitate further knowledge, heightened awareness and enlightenment.

Linkage between hexagonal crystals, DNA double helix, the cosmic serpent and indigenous peoples from various parts of the world is explainable when Narby (1999, p. 64) provides two illustrations from Reichel-Dolmatoff. In both instances the cosmic serpent is being led by a hexagonal crystal or is following a hexagonal crystal. In one illustration the serpent is lying in the fissure which separates both brain hemispheres. Outside the brain area there is a hexagonal crystal. If this drawing were to be depicted looking down onto the top of the head the crystal would occupy the space of what is commonly known as the third eye.

Atlantis

I am alluding to the, perhaps not so mystical, references made over the last one hundred years to the lost continents of Atlantis and Lemuria. Renowned psychic Edgar Cayce, early in the twentieth century, awakened worldwide interest in the possibility that there have indeed been civilisations more advanced than our present one and these pre-date conventional findings by thousands of years. In explaining energy sources used by the Atlanteans Cayce (2007, pp. 43-45) refers to a 'firestone' or crystal which was erected in such a way as to attract either the sun's rays or energy from the stars. Cayce elaborates further that these power sources were also used in a medical sense as a type of laser, enabling sophisticated surgical procedures to be performed.

Edgar Cayce died in 1945 leaving behind believers and sceptics, many of the latter being conventional scholars who seem to have an inbuilt fear of any discovery which may threaten their ideas of history. Fortunately in the second half of the twentieth century there were expeditions mounted which would suggest there is evidence which may substantiate Cayce's readings (Joseph, 2006, pp. 291-320).

Ayurveda

Western interest in crystals no doubt evolved from eastern and indigenous philosophies, Ayurveda being the most likely. The word itself means the science of life and embraces all aspects of healing including crystals. India is the home of Ayurveda where it has been widely in use for over five thousand years. To elaborate further on Ayurvedic philosophy here would do little or nothing to enhance one's knowledge; the reader is encouraged to seek out literature which offers insight at a deeper level of this most wholistic of all sciences.

Knowledge and use of gems and crystals within Ayurveda is one part of its system of healing. The sacred books of India teach that the seven cosmic rays are shown to us through the rainbow and gems are their physical or tangible manifestation. The origins of all life are contained within the seven cosmic rays including man, animals, trees, plants, rivers, mountains and rocks. Gems, it is believed, are a concentration of the colours which originated in the cosmic rays. It may be said that the rays are the life force or energy which flows through all things. The rainbow is a reminder to us of the perfection and omniscience of creation.

In themselves, rainbows are never permanent; they come and go but the content is unchanging. The colours are the same: violet, indigo, blue, green, yellow, orange and red. These are the cosmic rays. Ayurveda uses gemstones which are specific to these colours: sapphire equals violet; diamond equals indigo; moonstone or topaz equals blue; emerald equals green; coral equals yellow; pearl equals orange; ruby equals red.

When looking at these gems with the naked eye some of them will appear as different colours. For example, pearl does not appear to be orange any more than diamond appears indigo. The reason for this is the naked eye is unable to attune to certain colours. If we were to look at the gem through a prism (a transparent piece of triangular glass with refracting surfaces), the cosmic colours would be observed. Everything that is created when viewed through a prism will display the colours of the rainbow. Bhattacharya (1985, pp. 2-3) explains that something as simple as a pencil line drawn on a piece of paper when viewed through a prism will reveal the seven colours of the rainbow or the seven cosmic rays. Everything created by man or the Creator will reveal the seven cosmic rays.

Metals are also used in Ayurveda and are known to possess powerful healing energy. Copper, iron, silver, gold, tin and even mercury are all used. Preparation of these metals is specific as some contain toxic impurities. The process involves treatment with oil, milk, cow's urine, ghee and buttermilk. Treatment with gemstones, however, does not involve complicated preparation: place the stone(s) in water overnight and drink the water the following day (Lad, 1985, pp. 142-144). It is worth noting that when the water is drunk, not only is the mineral content being taken in to the body but also the ray component as well, which is the life force.

Australian Aborigines

Whether or not one believes in the existence of Lemuria and Atlantis, amongst more recent cultures of the Australian Aborigine and North and South American Indians the cultural practices which include the combined use of water and crystals would indicate knowledge of the power which emanates from such a combination. North American Indians use jasper, (trigonal/ hexagonal) referred to as the 'Rainbringer', for the purpose of rain making. Turquoise, while not a hexagonal crystal, does contain water and in combination with coral is also used as a rain maker by Navajo Indians. Coral, a word derived from the Greek, means Daughter of the Sea (Cunningham, 2002, p. 100, p. 172).

Could it be that the indigenous peoples have an innate knowledge which non-indigenous people are only just beginning to recognise – the use of trigonal/hexagonal jasper for attracting 'good' water. Is it just coincidence that some indigenous people use quartz crystal, also hexagonal, which is regarded by them as 'solidified water', for attracting rain; using the law of "Like Attracts Like"?

The same could be said for turquoise with water making up part of its chemistry. I maintain such examples serve to illustrate that the present day concept of maintaining health has deviated into a complicated maze from which it will be difficult to emerge. I contend that 'good' water (hexagonal) enhanced by a crystal (hexagonal) acts to boost the immune system. It does not necessarily cure illness, however, if the immune system is healthy, the human body is in a better position to ward off illness.

Use of crystals within the Australian Aboriginal culture offers a unique insight into the contribution crystals make in the psychic and spiritual world of the medicine man or smart man. Author A. P. Elkin (1994) travelled extensively across Australia in his role of Anthropologist with the University of Sydney and gained some insight into the initiatory practices of the medicine man. Apart from certain areas of the Kimberley region in North West Australia where pearl shell was widely used in the initiation ceremony, elsewhere Elkin observed that use of quartz crystals was an integral part of the initiation and they were regarded as being extremely powerful and of significance in the work of the medicine man. Small stones also formed part of the medicine man's equipment as did pieces of cord. The cord was vital as it enabled him to reach the sky for consultation with the spirits. Narby (1999, p. 63) discusses the similarities of shamans around the world with respect to ladders, ropes, vines and bridges. Lawlor (1991, pp. 374-375) describes the practice of inserting quartz crystals into the body as being part of many initiations amongst Aboriginal people. It is symbolic of the transformation of matter into an omnipotent, omniscient channel of cosmic energy. Quartz crystals with internal fractures are sought after as they produce brilliant rainbow coloured light refractions which signal to the Aboriginal medicine man that the crystal resonates with the primal energies of the Rainbow Serpent.

Michael Harner (1990, p. 140) refers to quartz crystals as "solidified light" involved in the process of "enlightenment". This concept is likened to the third eye. Aboriginal medicine men involved in initiations had a quartz crystal "sung" into their forehead thus enabling them to see in another dimension.

Australian Aborigines, isolated from the rest of the world for more than forty thousand years, refer to the Rainbow Snake in association with the power of quartz crystals. Several indigenous groups from South America refer to the cosmic serpent or sacred anaconda and the associated role of quartz crystals.

South American Indians

Within the culture of one such group, known as the Desana, use of quartz crystal is an integral part of the culture. Desana people are from the Vaupes region which lies within the Colombian northwest Amazon. The banks of the Vaupes River, one of the major rivers of that region, are home to some of the Desana while the remainder live in the Papuri River basin. Considered to be little known the Desana are one group who offer insight into the power of quartz crystal as part of their culture (Reichel-Dolmatoff, 1971, p. 5).

To the Desana people of South America the world was seen as a giant disk that was the dwelling place of the all-powerful Sun as well as being the home of men and animals. Yellow was the colour of the Sun, representing power, while red, representing blood and fecundity, was the colour assigned to men and animals. High above the earth, Sun created the Milky Way, the place where strong winds surge through. Its colour is blue: intermediate colour between Sun's yellow power and the red of the earth. For this reason it is considered dangerous because it is there that people can establish contact with the invisible world and the spirits (Reichel-Dolmatoff, 1971, pp. 24-25). The Sun's power was all-creative, all powerful. Father Sun created the birds, plants and animals causing the power of his yellow light to penetrate everything in order to create the world. Sun had a daughter whom he sent to Earth for the purpose of teaching people how to live well (Reichel-Dolmatoff, 1971, p. 35).

Reichel-Dolmatoff (1971, p. 48) explains the importance of yellow as understood by the Desana. Sun's yellow colour symbolises semen. Regardless of a particular shade of yellow, whether it be light, dark, opaque – all are associated with semen. The colour yellow is expressed in several ways, three of which are in liquid form; honey, saliva and semen. In a mineralised form, yellow becomes quartz crystal. All of these substances represent the fertilising power (or energy) of the Sun and as a consequence are of significance in Shamanistic practices.

Desana men of high status wear a yellow or white quartz cylinder suspended from their neck. Small pieces of quartz are significant and are placed inside gourd rattles as well as stored in their purse. The quartz cylinder is regarded as the "penis of the sun". Lightning is regarded as the ejaculation of the Sun; it can be for good or evil. When lightning has struck the wise-man looks for and collects pieces of scattered crystal so that they do not cause illness. These fragments of crystal are perceived as being dangerous seminal matter and as such need to be neutralised (Reichel-Dolmatoff, 1971, p. 129). These same quartz crystals having the power to cause illness are sometimes the means by which treatment is meted out to an offender.

Overall it is the quartz crystal symbolising the all-powerful sun that the Desana acknowledge, such power being the arbiter of birth and death.

Tibet
Tibetan physicians used gemstones, including quartz crystals, to make water potions for their patients (Tompkins & Bird, 1992, p. 104).

Hildegard of Bingen

More recently than the Atlantean period or Ayurveda is the work of a German nun Hildegard of Bingen (born 1098). In 1151 Hildegard established her own convent based on the Benedictine Order, at Rupertsberg near Bingen. According to Strehlow and Hertzka (1988, p. 141) the convent became a pivotal centre in Europe which was visited by thousands seeking advice from this remarkable woman. Hildegard had no formal training in medicine; her expertise lay in her philosophy of living and an extensive knowledge of plants and minerals. A deeply spiritual woman she taught that our own inner wisdom is far more advanced and highly developed than orthodox forms of study. In today's world Hildegard would be described as a wholistic practitioner, a person who embraced healing on all levels. Hildegard's extensive knowledge of foods, herbs and minerals provided the earthly requirements of her ministry but the core of her teaching, firmly embedded, was her mystical relationship with the Creator.

Author and herbalist Frawley (in Strehlow & Hertzka, 1988) suggests that Hildegard has indicated the return path we, as healers, need to take. True healing is about contacting the inner consciousness, the intuition; it is not an intellectual process. Since the publication of *Hildegard of Bingen's Medicine* in 1988, more than two decades ago, there has been a gradual dissemination of Hildegard's wisdom amongst practitioners of natural medicine. Several crystal remedies which Hildegard used eight hundred years ago are presented here.

A word of caution is issued regarding blue sapphire and lapis lazuli. According to Simmons and Ahsian (2005, p. 340) blue sapphire (Al_2O_3) was used by Hildegard for conjunctivitis and is used today for general eye problems. However, during the middle ages and further back in antiquity, what is recognised today as lapis lazuli was known as sapphire. The word originated from "sapphirus", a term used by ancient Greeks and Romans. It was not until the middle ages that the mineral became known as lapis lazuli (Simmons & Ahsian, 2005, p. 227; Schumann, 2009, p. 102). Lapis lazuli has a mineral component of $(NaCa)_8[(SiO_4SCl)_21(AlSi)_4]$. Whether the blue sapphire referred to by Hildegard was indeed sapphire or lapis lazuli as some researchers have indicated, is debatable. Gold was widely used for arthritis, gout, stomach catarrh (chronic build-up of mucous) and the immune system.

Modern literature records that skin cancer, arthritis, blood disorders, rheumatism, paralysis, heart disease and enhancement of mental faculties benefit from gold treatment. Hildegard's treatment involved powdered gold used in a gold paste and gold cookies. Gold treatment used in the first half of the twentieth century involved gold injections for the treatment of rheumatoid arthritis. Due to the cost involved and the increasing power of pharmaceutical companies gold treatment was not favoured by allopathic medicine. Gold is an element, Au.

Sardonyx was a mineral used by Hildegard to strengthen the five senses and the intellect. Today sardonyx is used in healing of lungs and bones as well as fluid regulation and cell metabolism. Other uses are strengthening of immunity and re-sensitising of sensory organs. The chemical composition of sardonyx is SiO_2 (Silicon).

Emerald is another mineral used by Hildegard in much the same way as it is used today. "Debility and infirmity" are changed back to health (Strehlow & Hertz, 1988, p. 38). We use emerald today for recovery from illness, treatment of sinus, lungs, heart, muscles and as an aid in liver detoxification. The chemical composition of emerald is $Be_3(AlCr)_2Si_6O_{18}$.

We arrive at the future on the stepping stones of the past.

Much more recently, science has revealed the physical and chemical properties of crystals and the roles they play in the information technology industry (Lawlor, 1991, p. 375). Recognition of the crystal's capacity to store, focus and transmit energy is now legend within the communication/information industry.

MINERALS

There are thousands of different minerals within the Earth's crust, however only a few are prevalent. The rest are found in minuscule amounts and become visible when there is a specific concentration caused by a marked geological formation. Mineral crystals form in several different ways. Formation occurs through heating and cooling, dissolving of chemicals in watery liquids, chemical alteration of existing minerals, and remaking of rock.

The selection of minerals for this book has been made for their use within the human body.

The following table from Haas' *Staying Healthy with Nutrition* (1992, p. 155), lists the elemental composition of the human body in order of quantity contained therein, that is, the body contains more oxygen than the other elements and chromium appears in the least amount.

Elemental composition of the human body

Macrominerals	*Key Function*
Oxygen	Cell and tissue respiration, water
Carbon	Protoplasm
Hydrogen	Water, tissue
Nitrogen	Protein tissue
Calcium	Bone and teeth
Phosphorus	Bone and teeth
Sulphur	Amino acids, hair and skin
Chloride	Electrolyte
Sodium	Extracellular electrolyte
Magnesium	Metabolic electrolyte
Silicon	Connective tissue
Microminerals	
Iron	Haemoglobin, oxygen carrier
Fluoride	Bones and teeth
Zinc	Metallo-enzymes
Strontium	Bone integrity
Copper	Enzyme cofactor
Cobalt	Vitamin B_{12} core
Vanadium	Lipid metabolism
Iodine	Thyroid hormones
Tin	Unknown
Selenium	Enzyme, antioxidant, detoxification
Manganese	Metallo-enzymes
Nickel	Unknown
Molybdenum	Enzyme cofactor
Chromium	Glucose tolerance factor

As more research is conducted on the use of minerals therapeutically, it could well be that more minerals will be added to the list of those found in the human body.

Already minerals such as boron and lithium are being used medicinally. What are referred to as toxic metals are also found in the body as naturally occurring elements – needless to say in miniscule amounts. These are aluminium, cadmium, mercury and lead. Through our mining practices these heavy metals have been removed from the earth and are used in industry hence the proliferation of heavy metal poisoning. Upsetting the balance and interfering with Nature has dire consequences, not only for humans but also the planet.

Arsenic is another element which is present in the human body although in micro amounts (Haas, 1992, p. 239).

> "Despite arsenic's reputation as a poison, it actually has fairly low toxicity in comparison with some other metals, although with chronic exposure there is some concern about arsenic's effect on chromosomes and its carcinogenicity. In fact, arsenic may even be essential and functional in humans in very small amounts."

Walters (2006, p. 118) also considers the importance of arsenic.

> "The mineral is a profound requirement for living systems just the same. In pregnancy, the arsenic level goes up tenfold if the child is to be a boy. Without such a level, spontaneous abortion becomes the inevitable result. Pine needles crushed and steeped as a tea in boiling water deliver enough arsenic to prevent miscarriage."

MINERALS - Helpful information

The minerals listed in the *Elemental Composition of the Human Body* table are contained within the fifty-two minerals/gems outlined in the Chemical Content Table (Appendix 4). Each mineral is vital for the correct functioning of the body. I do not intend to provide in-depth information on each but rather (with reference to Koch, 1993 and Haas, 1992) the following information about elements in the human body and how each forms an integral part of the whole.

CALCIUM – (Ca)
- The human body is composed of more calcium than any other mineral.
- Needed for skeletal muscle contraction.
- Essential for bones and teeth.
- Required for smooth functioning of the heart muscle.
- Needed for the involuntary muscular movements of the intestines.
- Vitamins A, C and D along with magnesium and phosphorous are necessary for the efficient functioning of calcium.
- Calcium is stored in the bones.
- Continual intake of sugars severely affects the calcium/phosphorous balance in the body. This gives rise to tooth decay, bone deformation in children and cancer sores.
- Excess alcohol and chocolate cause loss of calcium from the body.
- The use of table salt depletes the calcium levels.
- Cells and tissues of the body are enhanced by calcium which provides elasticity and firmness.
- Calcium is involved in the coagulation of the blood.
- The parathyroid and thyroid glands together control the level of calcium in the blood, to maintain blood pH of 7.4.
- Contained in nuts, legumes, most vegetables, dried peaches and apricots, dairy.

CHLORINE / CHLORIDE (Cl)
- Chloride works with sodium and water in the distribution of body fluids.
- An important ingredient in hydrochloric acid – the main digestive aid.
- Useful in assisting the liver with clearance of waste products.
- Chloride foods assist in regulation of acid-alkaline levels of blood.
- Assists in elimination of toxins and blood purification.
- Obtained readily from table salt.
- Foods which contain a good source of chloride are tomatoes, celery, lettuce, chives, rye and kelp.

CHROME (Cr)
- Plays a vital role in regulation of carbohydrates.
- Chromium is not easily absorbed and much that is absorbed is eliminated by faeces.
- Stored in fat, brain, muscles, spleen, kidneys and testes as well as other parts of the body.
- Tissue levels of chromium tend to decrease with age and this has been linked to adult-onset diabetes, which has increased more than sixfold over the last fifty years.
- Chromium reduces sugar cravings.
- Soil deficiency and food refining are possibly contributing factors to the increase in adult on-set diabetes.
- Chromium deficiency can also cause problems such as anxiety and fatigue as well as issues with blood sugar metabolism.
- Provided adequate amounts of chromium are in the soil, fresh fruit and vegetables are a good source of supply.

COBALT (Co)
- Cobalt is available as part of Vitamin B12.
- As part of Vitamin B12 cobalt is not readily absorbed from the digestive tract.
- Storage occurs in liver, kidney, spleen, pancreas and red blood cells.
- Deficiency can lead to pernicious anaemia and nerve damage.
- Soils are becoming deficient in cobalt, reducing further the levels found in plant foods.
- Vegetarians need to be concerned about getting sufficient levels of cobalt and Vitamin B12.
- Cobalt is supplied by fish and other sea foods.

COPPER (Cu)
- Copper forms part of every body tissue.
- Copper toxicity and zinc deficiency have increased since zinc water pipes were phased out and copper water pipes introduced.
- Half the copper content of the body is in the bone structure.
- Copper supplied by natural foods is absorbed into the bloodstream soon after digestion. Storage takes place in the kidneys, liver, heart, brain, bones and muscles.
- Important in the formation of haemoglobin.
- Birth control pills are linked to increases in serum copper.
- Deficiency of copper is linked with iron deficiency.
- Excess copper can interfere with zinc absorption.
- Best sources are nuts and seeds, olives, dates, apples, apricots.

FLUORIDE / FLUORINE (F)
- Fluorine is found in every bone.
- Assists in maintaining the function of the iris of the eye.
- Protects teeth enamel.
- Deficiency of fluorine may lead to impaired liver, kidney and heart function.
- Organic fluorine contains calcium fluorine but fluoridated water lacks the calcium content.
- Fluoridated water contains sodium fluoride and if taken in excess may have a toxic effect on the body.
- Organic fluorine from foods reduces acidity of the mouth which deters tooth decay.
- Barley, corn, wheat, fresh fruit, fresh vegetables, garlic, rice and oats are all sources of organic fluorine.

IODINE (I)
- Essential for regulation of thyroid gland and the production of thyroid hormones.
- Twenty per cent of iodine is stored in the thyroid gland.
- Assists the body to burn off excess fat.
- Deficiencies in iodine may lead to hardening of the arteries, irritability, obesity, rapid pulse, nervousness, poor concentration and problems with the thyroid gland.
- As well as in the thyroid gland, iodine is in skin and bones.
- Iodine needs the mineral selenium to make the thyroid gland work effectively.
- Best sources of iodine are to be found in fish, shellfish and kelp.

Note: Iodine is a mineral (iodarygyrite) but is not available for sale. The most widely used sources are contained in fish, shellfish and seaweed.

IRON (Fe)
- Oxygen requires iron for transfer within the body.
- Women require additional iron especially during menstruation.
- Iron deficiency may be connected to dizziness, fainting, muscle fatigue, difficulty breathing.
- The primary function of iron is the formation of haemoglobin in the body.
- Use of laxatives may contribute to loss of iron from the body.
- Too much tea and coffee affect absorption of iron.
- For the iron content of food to be adequately dissolved and absorbed a regular supply of B Vitamins is required by stomach hydrochloric acid.
- Copper and calcium are also required for effective iron absorption.
- Found in meat, rice bran, wheat bran, pumpkin seeds, sesame seeds, beans, peas, lentils, prunes, raisins, cashew nuts.

MAGNESIUM (Mg)
- Is a natural tranquilliser – relaxes muscles.
- Protects the heart.
- Helps to prevent kidney stones.
- Assists in treatment of pre-menstrual syndrome.
- Assists with reducing anxiety and insomnia.
- Acts to relieve muscle cramps.
- Acts to nourish the brain and spinal cord.
- The body does not store magnesium for long periods, therefore it must be obtained on a regular basis.
- Diabetics require a regular intake.
- Deficiency can be caused by alcohol consumption.
- Some food sources: almonds, brazil nuts, hazel nuts, sesame seeds, rice, lentils, dates, spinach.

MANGANESE (Mn)
- Important for digestion and utilisation of food, especially protein.
- Manganese foods assist with maintenance of normal blood sugar levels.
- Combined with iron and copper are necessary in formation of healthy red blood cells.
- Manganese and B group Vitamins stimulate transmission of impulses between nerves and muscles.
- Assists with hormone production – both male and female.
- Necessary for the regulation of menstrual cycles.
- Helps to promote good memory.
- Some food sources: kidney beans, lima beans, brazil nuts, almonds, coconut, walnuts, grapes, beetroot, parsley.

MOLYBDENUM (Mo)
- Considered to be an essential trace element.
- Primary source is from food, however it is commonly lacking in soils, so a deficiency in humans is quite common.
- Molybdenum is found mainly in adrenal glands, bones, skin, liver and kidneys but traces are found in all tissue.
- Has a vital role in uric acid formation and iron utilisation; as well it has importance in carbohydrate metabolism.
- The brain and nervous system require molybdenum.
- Depending on levels in soil, legumes and green leafy vegetables are a good source of molybdenum.

NICKEL (Ni)
- Is contained in the body but to date its uses are unknown.
- Does not seem to be concentrated in any particular tissues.
- Nickel is not well absorbed from the gastrointestinal tract.
- Most is eliminated through faeces, some in urine and a little in perspiration.
- Toxicity is of some concern, not from nickel found in foods but from nickel carbonyl, a cancer-causing gas from car exhaust, cigarette smoke and some industrial wastes.
- Allergic dermatitis is sometimes caused from nickel products such as jewellery, dental materials, contact with scissors, pins and some kitchen appliances.
- Foods containing nickel are: most beans, walnuts and hazelnuts. Oats, buckwheat, barley and corn also contain nickel.

PHOSPHOROUS (P)
- Essential for the repair and maintenance of the nervous system.
- Assists in transportation of fatty acids around the body.
- Forms part of every living cell in the body.
- Phosphorous foods assist in utilisation of fats, protein and carbohydrate.
- Foods containing phosphorous assist in reducing the likelihood of cancerous tissue formation.
- Vitamin D is essential for correct phosphorous balance.
- Needed for correct functioning of calcium along with magnesium and Vitamins A, C and D.
- In balance with calcium, tooth and gum problems may be alleviated.
- Phosphorous deficiency may be responsible for weakness, weight loss, stiff joints, anxiety, irritability.
- Found in most nuts, green vegetables, most fruits, meats, fish, cheese, eggs.

POTASSIUM (K)
- Potassium is essential for repair and health of all the body's muscles.
- In combination with sodium, blocks hardening of the arteries.
- Cancer cells cannot live in a potassium rich environment.
- Aids in balancing the pH levels in the body.
- Assists in liver repair.
- Necessary for formation of glycogen. Glycogen converts to glucose in the liver. Glucose is an energy source.
- Potassium is the most capable healing mineral in the body.
- Assists in the regulation of water balance within the body.
- Excessive alcohol and caffeine cause a potassium deficiency.
- Potassium is easily destroyed by cooking.
- Some food sources: root vegetables, dates, figs, most nuts, nectarines, apricots, avocado, berry fruits, leafy green vegetables.

SELENIUM (Se)
- Works with Vitamin E to stimulate the immune system.
- Most of the selenium in our body is in the liver, kidneys and pancreas.
- In men selenium is also found in the testes and seminal vesicles.
- Found to have an anti-carcinogenic effect.
- Works with Vitamin E to protect tissues and cell membranes.
- Assists in prevention of heart disease.
- Reduces the incidence of certain cancers.
- Good sources of selenium are found in barley, oats, wheat, rice, garlic, onions, broccoli and tomatoes, provided the soil in which they are grown contains sufficient amounts of selenium.

SILICON (Si)
- Essential for hair, skin and teeth.
- Acts as a cleanser, essential for proper blood circulation.
- Silicon is one of the most plentiful minerals found in soil.
- Found in bone, blood vessels, cartilage and tendons.
- Protects against nervous exhaustion, poor vision, mental fatigue, baldness.
- Deficiency of silicon related to arthritic conditions.
- Some food sources: lettuce, cucumber, cabbage, rye, soy bean, avocado, oranges, walnuts, peanuts, almonds, sunflower seeds.

SODIUM (Na)
- Sodium chloride (table salt) taken in excess contributes towards heart attacks, hardening of the arteries, cramps and strokes. (This is an inorganic form of sodium.)
- Sodium is of value when it is in an organic form such as celery, carrots, kelp, olives, parsley, raisins, sunflower seeds, broccoli, onions. Sea Salt and Himalayan Pink Salt provide an organic form of sodium.
- Organic sodium is good for skin, hair and eyes.
- Natural food does not have a high sodium (salt) content.
- Processed food contains high salt content which can lead to heart problems.
- Deficiency of organic sodium leads to imbalance of calcium and phosphorous.

STRONTIUM (Sr)
- Ninety-nine per cent of strontium is stored in bones and teeth.
- There is a close resemblance between calcium and strontium. Because of this strontium can displace calcium.
- Strontium content varies in our diet because it is available via the soil.
- Currently being explored for regeneration of bone cells.
- This form of strontium is not Radioactive Strontium (Sr_{90}) which is a by-product of nuclear fission.

SULPHUR (S)
- Highest protein absorption requires sulphur.
- Sulphur acts as an antiseptic and cleanser on the digestive system.
- Maintains normal fluid level of bile and pancreatic juices.
- Sulphur and phosphorous in balance, act as a blood purifier.
- Sulphur forms part of insulin structure which is required for carbohydrate metabolism.
- Correct oxygen balance requires sulphur.
- Sulphur is essential for normal function of heart muscles.
- Acts to prevent accumulation of waste toxins within the body.
- Essential for the formation of collagen, a substance found in connective tissue.
- Food sources: cabbage, Brussels sprouts, water cress, cauliflower, kelp, spinach, leek, radish, avocado, hazel nuts.

TIN (Sn)
- Present in the soil and in foods.
- Canned food sometimes adds tin to food.
- There is no known use for tin in the body.
- Tin is used as a fluoride carrier in toothpaste.
- Considered as mildly toxic – it is best to avoid.

VANADIUM (V)
- Vanadium is used for regulation of blood circulation.
- Storage in the body is mainly in fat and it is involved in fat metabolism.
- Helps some people in reduction of cholesterol.
- Since the late nineteen fifties has been recommended in the treatment of diabetes and atherosclerosis.
- Vanadium has a role in calcium metabolism, reproduction and in growth.
- Sources of vanadium are to be found in fish and vegetable oils.

ZINC (Zn)
- Supports the immune system.
- Important for the thyroid gland, pancreas, and reproductive organs.
- Necessary for skin, hair and nails.
- Zinc is essential for health of the prostate gland.
- Refined carbohydrates (cake, white bread, biscuits) often lack zinc content.
- Regular consumption of alcohol may lead to zinc deficiency.
- Stress increases the need for zinc.
- Birth control pills increase levels of copper, thus reducing zinc levels.
- Some food sources: oysters, almonds, walnuts, rice, barley, sunflower seeds, pumpkin seeds, wheat bran, sesame seeds, onions, asparagus.

METAPHYSICAL AND OTHER USES OF CRYSTALS

(A brief sketch with thanks to Melody, 2008 and Hall, 2003 and 2009)

1. **Almandine** – Assists in opening the pathway between base and crown chakras. Helpful in absorption of iron in intestines and in treatment of the liver and pancreas.

2. **Amethyst** – One of the well-known and most beautiful of crystals, amethyst is found in a variety of colours ranging from lilac through to purple; such hues depending on varying amounts of iron contained therein. Uses of amethyst are many and embrace the mental, physical, spiritual and emotional aspects of life. It is regarded as one of the principal stones for meditation and acts to integrate and balance energy of the physical and non-physical bodies. Amethyst has the ability to impart strength and stability while at the same time affords peace and stillness. The use of amethyst by royalty and the church for millennia is testament to its spiritual significance.
At the physical level amethyst has been used in treatment of addictions; stimulating the sympathetic nervous system; balancing the endocrine system; digestive disorders; and disorders of the skin and skeletal systems.

3. **Analcime (Analcite)** – Acts to clear and stimulate the heart chakra. Assistance mentally, physically and emotionally when one is entering a period of change.
Has been used for fluid retention and to assist with atrophied muscles.

4. **Andradite** – Enhances masculine qualities of courage, stamina and strength. Has been used to assist with alignment of magnetic fields within the body.
Aids assimilation of calcium, iron and magnesium in the body.

5. **Apatite (Blue-Green)** – Used in treatment of panic and personality disorders. Has been used for development of psychic gifts, spiritual attunement, building up energy and increasing motivation.
Assists with glandular problems, organs and meridians. Aids in formation of new cells.

6. **Apophyllite (Green)** – A crystal that completely embraces all realms. Having a considerable water content, apophyllite is an effective conductor of energy. It is a mineral that is used as a protection against fire and at an energetic level is used in activation of the heart chakra. Overall, apophyllite has a calming effect on mind and emotions.
At a physical level this mineral assists the respiratory system, particularly in regard to asthma. It has also been of assistance with skin regeneration.

7. **Aqua Aura Quartz** – This mineral has been embellished with pure gold which is not able to be removed by rubbing or scraping, and therefore has taken on the characteristics of gold as well as retaining the properties of quartz (Melody, 2008, p. 649).
Gold is regarded as a "master healer" and as such is an ideal mineral for cleansing the physical body and when used closely with another mineral offers stability to the other minerals. The uses of gold are many, both at a physical and non-physical level. These include treatment of depression; stabilising the emotions; and assistance with development and balancing of the heart chakra.
At the physical level gold has been used for the following: treatment of arthritis; blood disorders; skin cancer; rheumatism; eye problems; vascular dis-ease; spinal problems;

heart dis-ease; as well as absorption of vitamins and minerals; revitalisation of the endocrine system and the skeletal system.

The uses of quartz are many and varied and include stimulation of the immune system, protection of both the physical and emotional body, as well as acting as a balancer within all of the bodies – both physical and non-physical.

At the physical level quartz may be used to assist healing in any condition and is regarded as the most effective healing crystal of all. The ability of quartz crystal to store information, transmit and receive energy as well as focus does indeed make it one of the most powerful of all stones.

8 **Aquamarine** – One of the most powerful crystals for providing access to stored information. Regarded by many as a "stone of courage" aquamarine offers protective properties for the non-physical bodies as well as assisting in attunement to spiritual levels. A provider of emotional and intellectual stability, aquamarine offers a placid energy.
At a physical level this mineral assists with glandular problems; sore throats; teeth; poor vision; as well as acting as a general tonic.

9 **Barite** – A mineral which improves friendship and harmony as well as assisting in attainment of personal freedom from wants or needs of others, as well as from oneself. Sometimes barite acts powerfully to release trapped emotions followed by self-assurance and calm to one's inner self. Barite has been used to cleanse and stimulate the throat chakra.

10 **Bismuth** – A less widely used mineral but none the less one which offers characteristics conducive to healing. Bismuth has been used to aid the transition from the physical to the astral state as well as providing a certain comfort from feelings of emotional and spiritual isolation.
At a physical level bismuth has been used in the lessening of fevers and as an energy boost.

11 **Boleite** – This mineral has been used when working with the sixth chakra (third eye). It is used to enhance courage and/or action in an individual.
Physically boleite is used in treatment of throat ailments, including stabilisation of the thyroid gland, as well as microbial infections.

12 **Borax** – Has been used to cleanse chakras and to stimulate the crown chakra. The energy emanating from this mineral is claimed to be excellent for research purposes, acting to guide the researcher to sites of such information, for example libraries, newspapers, museums.
At a physical level borax has been used in treatment of infections; obesity; stomach upsets; and removal of toxins.

13 **Bornite (Peacock Rock)** – This mineral is a "stone of happiness" and acts to protect against negative energy. It is used in unification of the chakras but is also effective in energising individual chakras.
This mineral has demonstrated healing in both upper and lower parts of the body. Used above the waist bornite has assisted in the following: regulation of adrenaline flow; balancing acidity/alkalinity; assimilation of potassium; elimination of swelling and fever. Used below the waist bornite acts to increase circulation; stimulate adrenaline flow; and is useful in treatment of fevers and dry skin.

14 **Cassiterite** – This mineral offers positive energy in that it assists in dispelling prejudice, rejection, separation and disapproval. Also has been used in areas of astronomy and mathematics providing insight into "problem" areas.
At a physical level it has been used for hormonal imbalance and treatment of obesity.

15 **Cavansite** – Cavansite has been used for activating the intuition and as an aid for channelling.
At the physical level this mineral has been used for eye disorders; stabilising of the pulse; joint disorders; and assisting in the treatment of AIDS and Crohn's disease.

16 **Chabazite** – This mineral has many uses, some of which are described here. It has the ability to assist in reaching a meditative state and as a consequence chabazite allows for stillness of the mind. This state of mind in itself is conducive to creativity from which emanates the ability to manifest those ideas on the physical plane. Chabazite has been used to assist those with an addictive personality by enhancing the powers of stamina and determination.
At the physical level chabazite has been used in the treatment of obesity; muscular problems; and to assist in providing hormonal balance.

17 **Chalcopyrite** – Assists with enhancement of perceptive powers and aids as a connector so that information may be readily received.
At a physical level chalcopyrite has been used in cellular repair; reduction of inflammation; lessening of fevers; treatment of infectious diseases; and lung disorders.

18 **Chromite** – This mineral has the ability to neutralise unpleasant situations and lead one to an awareness of what could become potentially dangerous from a physical or emotional aspect.
Chromite has been used for eye disorders; healing at a cellular level; and aiding the absorption of Vitamins A and D.

19 **Chrysoprase** – One of the main uses of this mineral is assisting with chakra alignment, in particular energizing of the heart chakra. Chrysoprase assists with mental agility; stimulation of fluent speech; acceptance of self and others; as well as bringing about a sense of trust and security.
Physically this stone assists with heart problems; skin diseases; fungal infections; absorption of Vitamin C; and aids liver function.

20 **Citrine** – This mineral, belonging to the quartz family, has a wide variety of uses. It is one of the few crystals that never require cleansing as it does not retain negative energy. This attribute is of considerable importance when environmental protection is paramount. Other crystals may be cleansed by either placing them on top of or next to a citrine cluster. All aspects of "humanness" are enhanced by use of citrine due to its positive energy. Mentally and emotionally this crystal increases motivation, creatively, optimism, concentration, assists with emotional balance and the ability to rise above fear and depression. At an energetic level citrine is used for cleansing and activating the chakras.
Physically some of the uses for citrine involve energising and stimulating. Diseases such as chronic fatigue syndrome and glandular fever benefit from the use of this crystal. A sluggish digestive system, including the pancreas, is enhanced by the use of citrine. Improved blood circulation and immune function are also areas where citrine may play a role.

21 **Cobalt Aura Quartz (Siberian Blue Quartz)** – The healing properties of quartz in combination with cobalt blue offer a powerful tool in the treatment of throat disorders as well as stimulating the throat chakra. Any disorders which are regarded as inflammatory (such as gastritis, tonsillitis, arthritis, bursitis, hepatitis, bronchitis) would benefit from the use of cobalt aura quartz. The ability of quartz to transform, harmonise and focus energy plus the healing power of blue and its capacity to soothe and heal in all inflammatory illnesses makes this crystal a mineral "first aid kit".

22 **Cobaltite** – This stone has been used in the development and increase of creativity. The qualities of this mineral are that it can lead one to our innermost centre – the place where nothing is impossible and where true perfection resides.
At the physical level cobaltite has been used to treat cellular disorders; infectious illnesses; and dehydration.

23 **Covellite** – This mineral has been used where depression and despondency are involved. Covellite also assists in maintaining self-esteem at a positive level rather than a position of conceit; and also in the decision making process. Covellite encourages communication and positive analytic thought.
Physically this stone has been used as a protector against radiation and acts to stimulate cellular energy. It has been used in the treatment of eyes, ears, nose, mouth and sinuses as well as throat and fungal infections. Application of covellite on the body at the affected area has brought relief as well as bringing about a temperature increase of the mineral itself, indicating that progress is being achieved.

24 **Cuprite** – Acts to stimulate the base chakra and provides a grounding effect on the body. Cuprite has been found effective in the lessening of worry particularly in situations where one has no control.
Physically has been used to treat kidney and bladder disorders; also assists in treatment of altitude sickness; vertigo; arthritis; and rheumatism.

25 **Cyanotrichite** – This mineral has been used to increase one's planning capabilities and also acts to decrease argumentative speech connected to those planning capabilities.
At a physical level cyanotrichite has been used for the following health issues: muscular disorders; problems with the urogenital system; kidney disorders; and skin problems such as bruises, abrasions, burns, cuts and insect bites.

26 **Diaboleite** – One of the specific uses of this stone is that it offers continuity, through crystal-clear dreaming, between the physical and non-physical worlds.
At the physical level diaboleite assists with dysfunctions of the pineal gland and the brain. It has also been used for disorders affecting one's motor skills.

27 **Dioptase** – A significant stone for healing purposes, dioptase acts to energise all the chakras; balances the yin/yang energy; activates the memory of past lives; and raises one's consciousness.
Physically, dioptase has been used for cellular disorders; stimulation of T-cells; stress; tension; symptoms of Meniere's disease; and disorders of the heart and lungs.

28 **Emerald** – Said to be one of the stones used in the breast-plate of the High Priest, emerald is a stone of inspiration, integrity and unity. It is a mineral that encompasses physical, mental and emotional equilibrium and as a result brings about positive action. Emerald aids in the process of regeneration and recovery and is regarded as "a wisdom stone".

Physically emerald has been used to treat disorders of the heart, lungs, spine, muscles and eyes. It also is used as a liver detoxifier and in some instances as an antidote to poisons.

29. **Eudialyte** – This mineral appears to have resonance with those who have clairaudient tendencies. It has the ability to vary energies emanating from sound waves, blocking energy resulting from multiple transmissions and offering magnification of sound waves which are specific and well defined. Eudialyte has been used successfully in alpha and beta states as well as being involved in the opening and activation of the heart chakra. At the physical level eudialyte has been found useful in treatment of certain eye disorders relating to vision.

30. **Flame Aura Quartz** – This quartz crystal has been embellished with titanium and niobium, the molecules of which attach to the electric charge which surround the quartz crystal and are unable to be removed by scraping or rubbing (Melody, 2008, p. 651). The combination of colours created as the result of this allows for a wide range of healing leading to a state of well-being. This quartz crystal has been known to stimulate the flow of Kundalini from base to crown chakra, such movement leading to a balancing and alignment of the chakras.
At a physical level flame aura quartz has been used as: an immune stimulant; stabilisation of multiple sclerosis; mineral assimilation; a way to assist in release of toxins; and as a support in the treatment of bone cancer. Both titanium and niobium are used homoeopathically in the treatment of cancer (Scholten, 1996, p. 342). The structure of titanium is hexagonal (Scholten, 1996, p. 340).

31. **Fluorite (Fluorspar) (Rainbow)** – There are many different types of fluorite, each with its own distinctive qualities. Rainbow fluorite assists in transfer of information as well as cleansing, purifying and reorganising anything within the body that is not in a harmonious state. The energy of this mineral enhances self-sufficiency and protects against psychic attack.
At a physical level this mineral has been used for arterial pathways; deep tissue massage; emotional trauma; and disorders involving corpuscles.

32. **Gaspeite** – This stone allows one access to spirituality in one's daily life enabling the user to see beyond the veil of ignorance to a clarity of assurance. Gaspeite assists in one's ability to obtain and understand visions.
At a physical level this stone has been used to treat lung disorders and aids in the function of oxygen as it affects speech, thought patterns and memory.

33. **Grossularite** – Assists with relaxation and "going with the flow".
This stone is helpful in assimilation of Vitamin A and has been useful in treatment of arthritis, rheumatism, skin and mucus membranes.

34. **Gyrolite** – Uses of this stone are many and mainly deal with alignment in many forms. Uses of gyrolite in a non-physical capacity include stabilising the emotions; offering insight; strengthening the will; and in acupuncture to assist the well-being of the nervous system.
Physically, used in alignment of the body relating to the art of dancing; in the use of drums for meditation; laying on of stones; assisting in alignment of the spinal column; and maintenance of bone structure.

35 **Hemimorphite** – This stone acts to strengthen the ability to take responsibility for one's own actions and in so doing enables the concept of "know thyself" to reach fruition. It is very much a stone of personal evolution.
At a physical level it assists with blood disorders; cellular repair; pain reduction; treatment of herpes; and has been used as an aid to dietary support.

36 **Heulandite** – Assists in the lessening of ideas and conditions which cause distress such as jealousy, conceit, condescension. Allows for openness and a willingness towards lovingness thus encouraging compassion.
Physical uses include assisting with weight loss and dispersion of growths.

37 **Idocrase (aka. Vesuvianite)** – Has been used to remove negative thought patterns at a physical, emotional and mental level. It assists in bringing one closer and in touch with one's Higher Self and enhancement of one's creativity.
Physical uses include assisting with the assimilation of nutrients; lessening of skin problems; and increasing one's sense of smell.

38 **Jadeite** – Regarded by both Mayan and Aztec cultures as "a stone of magic" this stone assists in bringing about unity and cohesiveness within groups.
Physical uses include removal of pain (legs, hips); cellular and skeletal repair; and assistance with kidney problems. Jadeite has also been used for eye problems.

39 **Labradorite** – Regarded as a mystical and protective stone, labradorite protects one's aura as well as providing esoteric knowledge and guidance for initiation into the mysteries. This mineral provides calm to an overactive mind.
Physical uses include treatment of rheumatism; gout; colds; and lowering blood pressure. This mineral has also been used to relieve menstrual tension; regulation of metabolism; relief of stress; as well as treating eye disorders.

40 **Lazurite** – This stone enhances tranquillity and the spiritual nature of the user. It assists with intuition, thought transference and aids problem solving while in the dream state.
Physical uses include treatment of lung disorders; food allergies; and oxygenation of blood as part of disposal of toxins from the body. Other uses are minimising pain and inflammation.

41 **Magnesite** – Mentally and emotionally magnesite can assist when dealing with other people, their behaviour and/or addictions. At an energetic level this mineral is very useful in bringing a person to a level of peace in their meditation. Magnesite acts on the heart chakra and therefore assists greatly in the areas of unconditional love, compassion, tolerance, overcoming fears, impotence and other aspects of negativity.
As the name implies magnesite contains significant amounts of magnesium, consequently it may be used for any type of cramps such as stomach, intestinal, menstrual and any pain associated with kidney and gall stones.

42 **Mimetite** – This mineral offers the user the way to practical independence by alleviating the tendency to copy lifestyles, mannerisms, demeanour of others.
Physical uses include treatment of throat disorders; atrophy; and skeletal disorders.

43 **Mottramite** – For understanding the notion of "walk, don't run". Trust the process. Enjoying the result when the goal has been reached. Good for physical stress and labour – provides an adrenalin boost.

Physically used for adrenal glands, immune system, spleen, insomnia, lethargy. Good energy for use during convalescence. Energy acts to avoid relapse conditions.

44 **Nuummit** – This mineral is approximately four billion years old (the oldest living mineral) and was discovered in Greenland. May be used to activate, open and integrate the chakras. This mineral possesses many attributes including protective energy, enhancing memory and promoting recall.
At a physical level nuummit has been used for purification of the blood and kidneys; regulation of insulin; treatment of eye and throat disorders; and disorders of the central nervous system.

45 **Painite** – Use of this mineral provides emotional stability, and it assists in staying grounded.
At a physical level painite is useful in the treatment of blood disorders, removal of toxins, treatment of wounds, and reduction of inflammation. Painite has also been found effective in assisting the healing of bones and muscles.

46 **Phenakite** – The actions of this mineral appear to work at all levels; it is a good "all round" stone. The third eye is a primary point of activity whereby the energy generated flows throughout the body. However, the use of phenakite applied to any of the chakras produces an activating and cleansing energy which in turn acts upon the physical body.

47 **Pyrite** – This mineral has been used for stimulation of the third chakra. Pyrite is a "positive" stone, enhancing ability to establish new health patterns and creating positive energy. Wherever creativity is required, for example, art, sculpture, science, pyrite acts to stimulate the creative forces.
Physical uses include treatment of lung disorders; formation of cells; and energetic supporting of blood due to its iron content.

48 **Pyromorphite** – This stone has many facets, one in particular being the ability to enhance the energies of other stones that may be being used simultaneously. Pyromorphite is a crystal which offers the attributes of communication, teaching, healing, storage of information and the ability which allows for expansion of knowledge. This stone also integrates and balances the mental, physical, emotional and etheric bodies.
At a physical level pyromorphite has been used for the assimilation of the B vitamins; treatment of conditions involving the gums and connective tissue of the stomach; eye disorders; lung conditions; and providing healing energy to specific body systems and organs.

49 **Quartz Silver** – The silver and quartz properties are combined and produce powerful energies. The ability of quartz to transmit and receive energy as well as store and focus energy enables the silver component to be greatly enhanced. Use of this stone when astral travelling allows for a safe return to the earth plane.
At the physical level silver has been used in the elimination of toxins via the skin as well as at a cellular level. Used in the treatment of hepatitis and in the assimilation of vitamins B6, A and E.

50 **Rhodochrosite** – This mineral has often been referred to as a "stone of love and balance" and according to Melody (2008, p. 689) transmits a pulsating energy characteristic of the power of love. Hall (2003, p. 244) refers to rhodochrosite as being "an excellent stone for the heart and relationships." At the energy level this stone is ideal for the solar plexus

and base chakras assisting in the clearance of blocked energies. At the mental level rhodochrosite is stimulating and produces a positive, creative attitude.

Physically this stone has been used in the treatment of asthma and related respiratory problems. The circulatory and urinary systems benefit from the purification which rhodochrosite offers. It normalises blood pressure, stabilises heartbeat and dilates blood vessels.

51 **Rhodonite** – This stone is "a stone of power" assisting one to recognise where one's talents and energies lie and using those talents to enhance the gifts of others. It is a stone for activating the heart chakra.

Physical uses for rhodonite include treatment of emphysema; inflammation of joints; stomach ulcers; wounds; assisting bone growth; and gall bladder problems.

52 **Rutile** – Can be described as an "enhancer". It enhances learning, creative processes, accelerates intuition, consciousness.

Physically, rutile energetically assists the endocrine system and digestive system and is used to assist with addictions, for example tobacco, caffeine and food.

53 **Scapolite** – Offers "strength of purpose" in whatever endeavour one is seeking assistance, be it memory of past lives, writing a book, losing weight, doing away with bad habits, delving into one's psyche.

Depending on its colour, scapolite is used to treat the following:

Grey/White – assists in regulating brain chemistry.

Blue – eye problems including nerve imbalances.

Yellow – digestion: assimilation and elimination; kidney and gall bladder disorders.

Pink – all stress related disorders, particularly ulcers and digestive upsets.

54 **Selenite** – This mineral is used for attaining clarity of mind, ideal for meditation. It can also be used as a protector around the home, ensuring a harmonious atmosphere.

At a physical level selenite enhances the immune system and has shown benefit in treating cancer, tumours, and other disorders emanating from free radicals. It has also been used to treat spinal disorders.

55 **Smithsonite** – This stone provides calm and counteracts all negative energy. It offers clarity, compassion and support in all situations by relaxing the mind. There are several different colours of smithsonite – blue, pink, purple, green, yellow, white, grey and brown – each of which is specific to a particular chakra. Examples are pink which acts to stimulate the heart chakra and the release of old negative patterns; and white which when used on the crown chakra has the ability to bring energy from the other chakras into alignment thus promoting vitality and awareness.

At a physical level smithsonite assists in the balancing of both the endocrine and reproductive systems; the treatment of digestive disorders; osteoporosis; restoration of tone/elasticity to muscles and veins; as well as increasing physical energy.

56 **Sodalite** – This stone has a strong affinity with intuition and perception. It instils a passion for truth and drive for idealism. Sodalite is a rewarding stone for the mind as

it brings clarity thus removing mental confusion and any tendency toward intellectual enslavement.

At a physical level sodalite boosts the immune system; assists in treating effects of radiation; is useful for throat problems, for example, hoarseness, vocal cords; lowers blood pressure; and cools fevers.

57 **Sphalerite** – This stone is significant in its ability to "ground". It energises the first, second and third chakras providing vitality, willpower, zest for life and creativity.

Physical uses include support for the immune system; preventing "burn out"; treatment for eye disorders; and assistance with intake of nutrients.

58 **Strontianite** – This stone promotes the ability to be flexible and adaptable. It is a stone of strength acting to encourage growth in all aspects of living.

The physical uses of strontianite are diverse. It has been used to protect and energise house plants, within the home environs, as a protector from pests. Reduction of high acidity in the body; treatment of swelling; improvement of eyesight; and increasing vitality are some of strontianite's uses.

59 **Sturmanite** – This stone offers a variety of qualities relating to all aspects of being: mental, physical, emotional and spiritual. It is a useful stone when dealing with the solar plexus as it has the ability to clear, balance and stimulate. Sturmanite offers positive qualities relating to intellectual activities such as study, higher education, comprehension, teaching, religion and the study of languages. While this stone activates yin qualities it does also act to balance the yang properties and offers vitality to the physical body. It is a valuable stone for discernment, allowing a person to distinguish between negative and positive attitudes; an example would be recognition of frankness and deviousness.

At a physical level sturmanite has been used in treatment of the throat and digestive system, in particular the stomach, liver, gallbladder and pancreas. It has been used to support the adrenal glands, assist in the control of emotional disorders and aids the process of toxic elimination.

60 **Sugilite** – This stone activates all the chakras with its powerful purple ray. It is regarded as a "love stone" capable of bringing healing to all humankind. Sugilite is used with groups because of its ability to dispel conflict. This stone is beneficial to those who suffer from mental problems as it has the capability of "grounding" the person.

At a physical level sugilite is of benefit to cancer sufferers because of its ability to release emotional upheaval. It is of benefit as a pain reliever; useful for clearing headaches; and treatment of the nervous system.

61 **Topaz** – The energy of this mineral acts through the laws of manifestation and attraction. Also used in visualisation, meditation and projection.

There are many different coloured topazes each having different uses physically. For example, golden topaz is used in liver, gall bladder and endocrine disorders while purple topaz has been used for disorders such as autism, schizophrenia and other personality disorders.

62 **Tourmaline** – A mineral that has been used by Shamans of African, American, Indian and Aboriginal origins as a bringer of healing powers and protection. It has been used to bring balance to both brain hemispheres, energise all chakras and rebalance meridians. There are many different coloured tourmalines each with special healing attributes. Some of these are:

a. **Black Tourmaline** – acts to strengthen immune system, relieves pain, protects against all types of electromagnetic energy including cell phones and negative energies of all varieties, for example, psychic attacks, spells.

b. **Light Blue Tourmaline** – This colour acts to promote psychic awareness and visions, as well as encouraging harmony with the environment. This colour aids the immune system and lungs; assists with kidney and bladder disorders; aids thyroid imbalance; assists with insomnia; night sweats; bacterial infection; eye problems; and acts to soothe burns.

c. **Dark Blue Tourmaline** – Especially helpful with eye problems.

   Both light and dark blue tourmaline can be used anywhere there is congestion.

d. **Brown Tourmaline** – Ideal stone for grounding and clearing the aura. This stone may be used for family relationship problems. Physically is used for intestinal disorders and skin diseases.

e. **Green Tourmaline** – Strengthens the nervous system; treats disorders of the eyes, brain, immune system; relieves exhaustion; and is helpful in treatment of strained muscles. Green tourmaline opens the heart chakra thus promoting compassion, understanding, patience and creativity.

f. **Multi-coloured Tourmaline** – Because it contains all colours this stone brings body, mind, soul and spirit into harmony. At a physical level its uses are stimulation of the immune system and metabolism.

g. **Pink Tourmaline** – is used as an aphrodisiac in both the spiritual and material world and provides connection to wisdom and compassion.

h. Used at a physical level this stone assists in balancing endocrine dysfunctions, also treats lungs, skin and heart.

i. **Purple-Violet Tourmaline** – provides connection to base and heart chakras as well as unblocking the third eye chakra. Provides stimulation of the pineal gland.
In a physical sense this stone treats Alzheimer's and pollutant sensitivity.

j. **Red Tourmaline** – Because of its colour this stone increases stamina and endurance and offers vitality to the physical body. It assists in treating blood vessels, digestive system, heart and reproductive system.

k. **Yellow Tourmaline** – is used to stimulate the solar plexus and increases personal power.
At a physical level this stone treats liver, spleen, stomach, gallbladder and kidneys.

63 **Uvarovite** – This stone offers enrichment of tranquillity, solitude and calm and in so doing provides mental clarity.
Physical uses include treatment of kidney and bladder disorders; acidosis; lung disorders; and leukaemia.

64 **Vanadinite** – Assists in bringing order to one's life, defining goals and providing overall energy. At an energetic level it has been found useful in meditation, either to alleviate mind chatter or to facilitate awareness.
Physical uses include treatment of lung disorders; bladder dysfunction; and general states of exhaustion. As this mineral contains lead, if using ingestively it is advisable to make by the indirect method rather than placing the crystal directly into water.

65 **Willemite** – Metaphysically willemite is a useful stone for those who are at the beginning stages of their spiritual journey. It offers a calm, untroubled approach to what sometimes can be a time of uncertainty and apprehension. Willemite offers the user of this stone the ability to see themselves in a discerning way without being judgmental.
At the physical level this stone has been used in the treatment of fungal infection; hepatitis; dizziness; and disorders of the eyes affecting clarity of vision.

66 **Wulfenite** – A stone which balances the individual including those who have difficulty in accepting the shadow side of their nature. Wulfenite offers ease of movement from the physical through to all levels of being including movement through time.
While there are no recorded physical uses of wulfenite it does have power of rejuvenation and preservation.

67 **Xenotime** – This mineral has been used to illustrate that combined use of creativity and the Law of Creation will generate whatever is required, provided it is for the good of all. This stone offers strength of purpose when proceeding with one's goals.
Physical uses are: aiding the metabolising and absorption of phosphorous; and assisting with pH balance when there is excess alkalinity (yttrium is contained in xenotime and in recent years has been subject to research on Alzheimer's).

68 **Zincite** – Is a stone of power at both a physical and spiritual level. It assists in the removal of energy blocks thus permitting the vital life force to flow without hindrance. Examples of the power of this stone are: alleviates trauma; allays depression; promotes courage.
Physical uses are: treatment for skin and hair; prostate gland; menstrual and menopausal issues; and immune system (including Candida, auto-immune diseases, CFS and AIDS).

69 **Zircon** – This stone offers strength of purpose in achieving goals within the physical and spiritual realms. It is symbolic of purity and stability in every facet of one's life.
Some physical uses are: treatment of sciatic nerve disorders; treatment of vertigo; muscle repairs; and boosting bone stability.

MINERAL TINCTURES

There are several ways by which tinctures may be made, most favour the sunlight or moonlight method (see *Preparation of Mineral Tonics*). A point of difference occurs when establishing a time frame for the actual process. There are numerous choices, some of which are:

- In sunlight for twelve hours;
- In sunlight preferably for a whole day but at least three hours;
- In sunlight from four to twelve hours;
- In sunlight twenty-four hours followed by seven days in the sun or the new moon or the full moon followed by seven days in the dark;
- In the dark for a period of seven days and seven nights.

I have a theory based on personal experience. I have made and used tinctures based on the seven days and seven nights in the dark method. These tinctures were used for physical ailments only and the crystals were not programmed. At the time of making I had been working with magnetising of water and was able to prove that the longer the water was exposed to the magnet the more powerful it became. My theory is that during the seven days and seven nights in the dark process the medium (water) was increasing in strength and the exchange of energy was based on the transfer of more gross (physical) rates of vibration, that is, the mineral component, and as a consequence, physical ailments were being addressed.

With the sunlight or moonlight method for exchange the effect is the transfer of more subtle (or finer) rates of vibration. Emotional and spiritual vibrations are more subtle than gross physical vibrations. There are two components here:

- The energy emanating from the sun and the moon; and
- The length of time is considerably less.

The seven days and seven nights method is a total of one hundred and sixty-eight hours as against the twelve hours of the sunlight method. The twelve hours is one fourteenth or seven per cent of the time taken for the former method. I believe the tinctures I made using the longer method effectively treated physical ailments because the length of time allowed for the transfer of the mineral energies to occur. It was the energetic minerals that treated the physical ailments.

An example of this is the use of a lapis lazuli tincture. My client suffered from vertigo, throat and larynx problems and a compromised immune system. Lapis contains sodium, calcium, silica, oxygen, sulphur, chlorine and aluminium. The reduction of the vertigo occurred within twenty-four to thirty-six hours, and the remaining three symptoms cleared after two to three days.

Each crystal is unique as is each person. No doubt some crystals will release frequencies that deal with both physical and non-physical aspects of a person even though the crystal extraction occurs by the shorter method. It is worthwhile experimenting with the different times using a variety of crystals.

PREPARATION OF MINERAL TONICS

1. Select the mineral/crystal from which the mineral water is to be made.
2. Thoroughly cleanse the mineral/crystal before commencing the mineral water process. There are numerous ways by which this can be accomplished. Three of the most common are with the use of salt, however, water-soluble crystals should be cleansed in a more appropriate manner.
   - Cleanse the crystal by immersing it in seawater followed by energising in sunlight.
   - Immerse in sea salt added to water. Leave for a period of seven to twenty-four hours. Rinse in pure water (distilled water, spring water) and leave in sunshine to energise.
   - Soak crystal in prepared salt water from one to seven days. (Three tablespoons salt to one cup of distilled or spring water.) Use a clear glass container for this purpose and leave it in a sunny location. Rinse in pure water after the cleanse is complete.

   Alternatively, place the crystal in the moonlight during the night of a full moon.
3. Place the now-cleansed crystal in a clear glass container.
4. Place this container into a clear glass bowl containing distilled water, spring water or any other non-polluted water.
5. Place this in the sunlight, moonlight or both for twenty-one to twenty-four hours. I prefer to increase the time left in the sunlight/moonlight to seven days. To remove at the end of twenty-four hours, however, is sufficient and the resultant mineral water is effective for most uses.
6. To preserve what is called the Mother Elixir, add thirty per cent brandy to the mineral water and shake. For example, into sixty millilitres of mineral water add eighteen millilitres of brandy.
7. Store in capped amber bottles away from sunlight.
8. Store away from electrical fields such as refrigerators and microwave ovens.
9. Dosages would differ according to:
   - The mineral being taken – for example, referring to the *Elemental Composition of the Human Body* table, the dosage required for calcium would be far greater than that of chromium.
   - The age of the person requiring them.
   - See section on *Dosages*.

The safest method of preparation, known as the Indirect Method, is to place the cleansed crystal in a clear glass container and place that container into a clear glass bowl which holds distilled water or spring water. This method provides safety from any residual contamination which may remain after cleansing and also from use of any possible toxic minerals.

Please note, there are minerals which dissolve in water, two of which are borax and selenite, and can therefore be cleansed differently.

Cleansing of water soluble crystals
Crystals may be cleansed without water if necessary.
- One method is to "smudge" the crystal with herbs such as sage, myrrh or cedar.
- Another is to place the crystal requiring cleansing onto crystals such as blue kyanite, green obsidian, or citrine. Selenite is also a crystal which is used for cleansing.
- A third is to place the crystal in the moonlight during the night of a full moon.

DOSAGES

It is important to know that where dosages are concerned "one size does not fit all". The range may vary from three millilitres per dose to twenty millilitres per dose and this depends on the age and the weight of the person concerned.

It is usual to have two or three doses per day before a meal. Dilution with added water would lessen the strength not increase the strength. (Remember this is not a homoeopathic medicine you are dealing with.)

Children under twelve years of age require a child's dose. Children twelve years or over require an adult dose, depending on their weight.

Commencing at ten kilos and increasing by increments of five kilos children's dosages would be:

- 10 kg / 1.5 stone ⎫
- 15 kg / 2.35 stone ⎬ 2 – 4 years of age
- 20 kg / 3.14 stone ⎪
- 25 kg / 3.92 stone ⎭
- 30 kg / 4.7 stone ⎫
- 35 kg / 5.5 stone ⎬ 4 – 12 years of age
- 40 kg / 6.3 stone ⎭

Children from age two up to and including four years of age would start at three millilitres per dose within a weight range of between ten and twenty-five kilos.

Within the age range of four to twelve years and a weight range of twenty-five to forty kilos, dosage would be seven millilitres per dose.

Adult dosage within a weight range of forty to eighty kilos and age range between twelve to sixty years would have a dose of twenty millilitres per dose.

Adults over sixty years of age, no matter the weight, have a dose of fifteen millilitres per dose.

It would be considered inappropriate to administer to children up to the age of two years.

For adults whose weight is forty kilos and under start dosage at ten millilitres per dose for three weeks then increase to fifteen millilitres, then (depending on the individual's state of health) gradually, increase to twenty millilitres.

Note: The aforementioned dosages are based on the seven days and seven nights method and the person's physical ailments. If any of the sunlight/moonlight methods is used it would be appropriate to base the dosage on those applied to use of flower essences. It is also appropriate to dowse, muscle test and/or visualise the relevant dosage for individuals.

A less involved way of using crystals rather than the dosage method as outlined previously, is ingesting mineral-water as a drink. Select the crystal to be used and place it in a glass of 'good' magnetised water for two to three hours. This can then be drunk as required. I use this method more frequently than the dosage method because it relates more closely to the concept of 'good' water resonating with the body's cells. If a crystal is selected which is hexagonal, it too is adding to the resonance of the 'good' water (itself hexagonal) and the overall effect on the cells.

It needs to be remembered that each person is unique therefore the time allowed for the crystal to be immersed in the glass of water is individual. Please check this aspect before using a crystal in this way. Use of a pendulum or checking with kinesiology provides an answer to this question.

There have been situations where I have immersed a second crystal in the glass of water. Each time the second crystal has been a clear quartz crystal and this has added strength, energy and focus. It is not always possible to use a hexagonal crystal; the particular crystal you need to work with may be one of the other six crystal types. With the addition of the clear quartz crystal, hexagonal in itself, the strength, energy and focus of the water is enhanced.

APPENDIX 1 – The Crystals Chosen

Fifty-two crystals have been chosen primarily to address the needs of our physical and emotional health. Twenty-six crystals relate to the elemental composition of the human body. As the crystals depict an emotional component as well as the mineral content, fifty-two have been presented to offer a wider selection of a possible underlying emotional condition. Those chosen are:

Almandine	Cyanotrichite	Rutile
Analcime (Analcite)	Diaboleite	Scapolite
Andradite	Dioptase *	Selenite
Apatite *	Emerald *	Smithsonite *
Apophyllite	Fluorite (Fluorspar)	Sodalite
Aquamarine *	Gaspeite *	Sphalerite
Barite (Baryte)	Grossularite	Strontianite
Boleite	Hemimorphite	Sugilite *
Borax	Heulandite	Topaz
Bornite	Idocrase (Vesuvianite)	Tourmaline *
Cassiterite	Jadeite	Uvarovite
Cavansite	Labradorite *	Vanadinite *
Chalcopyrite	Lazurite	Wulfenite
Chromite	Mimetite *	Xenotime(y)
Chrysoprase *	Mottramite	Zincite *
Cobaltite	Nuummit	Zircon
Covellite *	Pyrite	
Cuprite	Rhodonite	

An additional seventeen hexagonal or trigonal/hexagonal crystals have been included for use with 'good' water which in itself is hexagonal. Some crystals in the fore-going fifty-two (marked with an asterisk) are also hexagonal or trigonal/hexagonal and can be used as well as the following:

Amethyst	Eudialyte	Pyromorphite
Aqua Aura Quartz (Enhanced with 24 carat gold)	Flame Aura Quartz (Enhanced with Titanium)	Quartz/Silver (Enhanced with Silver)
Bismuth (Treated sample)	Gyrolite	Rhodochrosite
Chabazite	Magnesite	Sturmanite
Citrine	Painite	Willemite
Cobalt Aura Quartz (Enhanced with Cobalt)	Phenakite	

APPENDIX 2 – The Pictures

The following photographs of the crystals used in this book are included for your information and, if desired, divination purposes.

Number This is chronologically given to the alphabetically arranged crystals and can be cross-referenced in the *Metaphysical and Other Uses of Crystals* section.

Crystal Name Fifty-two crystals (as per Appendix 1) are chosen for their elemental content. A further seventeen (as per part two of Appendix 1) are additionally chosen as hexagonal or trigonal/hexagonal crystals for their use with 'good' water. Refer to the *Chemical Content* tables (Appendix 3 & 4) for each crystal's system type, chemical formula and colours.

Crystal Shape The shape may be of assistance when crystals are used as a healing tool. Refer to the significance of this in *Numerology and Crystals*. The Numerology, or the power of numbers, is an aspect that can be traced to a time preceding Pythagoras and his School at Crotona, Circa 500 BCE.

1 **Almandine**
Shape: Cubic
Contains:
- Iron
- Aluminium
- Silicon
- Oxygen

2 **Amethyst**
Shape: Trigonal/Hexagonal
Contains:
- Silicon
- Oxygen

3 **Analcime**
Shape: Cubic
Contains:
- Sodium
- Aluminium
- Silicon
- Oxygen
- Hydrogen

4 **Andradite**
Shape: Cubic
Contains:
- Calcium
- Iron
- Silicon
- Oxygen

5 **Apatite**
Shape: Hexagonal
Contains:
- Calcium
- Fluorine
- Chlorine
- Phosphorus
- Oxygen

6 Apophyllite
Shape: Tetragonal
Contains:
- Potassium
- Fluorine
- Calcium
- Silicon
- Oxygen
- Hydrogen

7 Aqua Aura Quartz
Shape: Trigonal/Hexagonal
Contains:
- Silicon
- Oxygen
- Gold

8 Aquamarine
Shape: Hexagonal
Contains:
- Beryllium
- Aluminium
- Silicon
- Oxygen

9 Barite
Shape: Orthorhombic
Contains:
- Barium
- Sulphur
- Oxygen

10 Bismuth
Shape: Trigonal/Hexagonal
Contains:
- Bismuth

This sample has been treated by powdering the mineral and placing this into a boiling solution which results in reformation of crystals into various shapes.

11 Boleite
Shape: Tetragonal
Contains:
- Lead
- Copper
- Silver
- Chlorine
- Oxygen
- Hydrogen

12 Borax
Shape: Monoclinic
Contains:
- Sodium
- Boron
- Oxygen
- Hydrogen

13 Bornite (Peacock Rock)
Shape: Cubic
Contains:
- Copper
- Iron
- Sulphur

14 Cassiterite
Shape: Tetragonal
Contains:
- Tin
- Oxygen

15 Cavansite
Shape: Orthorhombic
Contains:
- Calcium
- Vanadium
- Oxygen
- Silicon

16 Chabazite
Shape: Trigonal/Hexagonal
Contains:
   Calcium   Oxygen
   Aluminium Hydrogen
   Silicon

17 Chalcopyrite
Shape: Tetragonal
Contains:
   Copper
   Iron
   Sulphur

18 Chromite
Shape: Cubic
Contains:
   Iron
   Chromium
   Oxygen

19 Chrysoprase
Shape: Trigonal/Hexagonal
Contains:
   Silicon
   Oxygen

20 Citrine
Shape: Trigonal/Hexagonal
Contains:
   Silicon
   Oxygen

21 **Cobalt Aura Quartz**
(Siberian Blue Quartz)
Shape: Trigonal/Hexagonal
Contains:
  Silicon
  Oxygen
  Cobalt

22 **Cobaltite**
Shape: Cubic
Contains:
  Cobalt
  Arsenic
  Sulphur

23 **Covellite**
Shape: Hexagonal
Contains:
  Copper
  Sulphur

24 **Cuprite**
Shape: Cubic
Contains:
  Copper
  Oxygen

25 **Cyanotrichite**
Shape: Orthorhombic
Contains:
  Copper   Oxygen
  Aluminium Hydrogen
  Sulphur

APPENDIX 2 – The Pictures

26 **Diaboleite**
Shape: Tetragonal
Contains:
Lead   Oxygen
Copper   Hydrogen
Chlorine

27 **Dioptase**
Shape: Trigonal/Hexagonal
Contains:
Copper
Silicon
Oxygen
Hydrogen

28 **Emerald**
Shape: Hexagonal
Contains:
Beryllium   Silicon
Aluminium   Oxygen
Chromium

29 **Eudialyte**
Shape: Trigonal/Hexagonal
Contains:
Sodium   Manganese   Oxygen
Calcium   Yttrium   Hydrogen
Cerium   Zirconium   Chlorine
Iron   Silicon

30 **Flame Aura Quartz**
Shape: Trigonal/Hexagonal
Contains:
Silicon
Oxygen

31   **Fluorite (Fluorspar) (Rainbow)**
   Shape: Cubic
   Contains:
       Calcium
       Fluorine

32   **Gaspeite**
   Shape: Trigonal/Hexagonal
   Contains:
       Nickel   Carbon
       Magnesium   Oxygen
       Iron

33   **Grossularite**
   Shape: Cubic
   Contains:
       Calcium
       Aluminium
       Silicon
       Oxygen

34   **Gyrolite**
   Shape: Trigonal/Hexagonal
   Contains:
       Sodium   Aluminium
       Calcium   Oxygen
       Silicon   Hydrogen

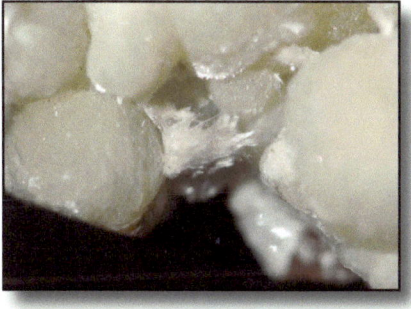

35   **Hemimorphite**
   Shape: Orthorhombic
   Contains:
       Zinc
       Silicon
       Oxygen
       Hydrogen

36 **Heulandite**
Shape: Monoclinic
  Contains:
  Calcium   Silicon
  Sodium   Sulphur
  Potassium Oxygen
  Aluminium Hydrogen

37 **Idocrase (aka Vesuvianite)**
Shape: Cubic
Contains:
  Calcium   Silicon
  Magnesium Oxygen
  Iron   Hydrogen
  Aluminium

38 **Jadeite**
Shape: Monoclinic
Contains:
  Sodium
  Aluminium
  Silicon
  Oxygen

39 **Labradorite**
Shape: Trigonal/Hexagonal
Contains:
  Calcium   Aluminium
  Sodium   Oxygen
  Silicon

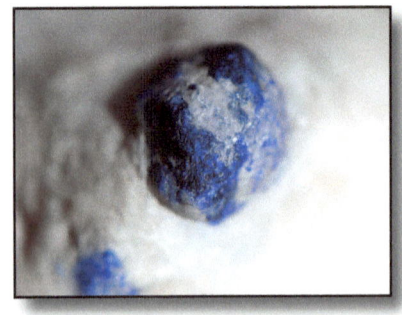

40 **Lazurite**
Shape: Cubic
Contains:
  Sodium   Oxygen
  Calcium   Sulphur
  Aluminium Chlorine
  Silicon

41 Magnesite
Shape: Trigonal/Hexagonal
Contains:
- Magnesium
- Carbon
- Oxygen

42 Mimetite
Shape: Hexagonal
Contains:
- Lead
- Arsenic
- Oxygen
- Chlorine

43 Mottramite
Shape: Orthorhombic
Contains:
- Lead
- Copper
- Vanadium
- Oxygen
- Hydrogen

44 Nuummit
Shape: Cubic
Contains:
- Magnesium
- Iron
- Aluminium
- Silicon
- Oxygen
- Hydrogen

45 Painite
Shape: Hexagonal
Contains:
- Calcium
- Zirconium
- Aluminium
- Oxygen
- Boron

46 **Phenakite**
Shape: Trigonal/Hexagonal
Contains:
Beryllium
Silicon
Oxygen

47 **Pyrite**
Shape: Cubic
Contains:
Iron
Sulphur

48 **Pyromorphite**
Shape: Hexagonal
Contains:
Lead
Phosphorus
Oxygen
Chlorine

49 **Quartz Silver**
Shape: Trigonal/Hexagonal
Contains:
Silicon
Oxygen
Silver

50 **Rhodochrosite**
Shape: Trigonal/Hexagonal
Contains:
Manganese
Carbon
Oxygen

51 Rhodonite
Shape: Triclinic
Contains:
Manganese   Calcium
Iron   Silicon
Magnesium   Oxygen

52 Rutile
Shape: Tetragonal
Contains:
Titanium
Oxygen

53 Scapolite
Shape: Tetragonal
Contains:
Calcium   Cobalt
Silicon   Sulphur
Aluminium   Chlorine
Oxygen

54 Selenite
Shape: Monoclinic
Contains:
Calcium
Sulphur
Oxygen
Hydrogen

55 Smithsonite
Shape: Trigonal/Hexagonal
Contains:
Zinc
Carbon
Oxygen

56 **Sodalite**
Shape: Cubic
Contains:
- Sodium
- Aluminium
- Silicon
- Oxygen
- Chlorine

57 **Sphalerite**
Shape: Cubic
Contains:
- Zinc
- Iron
- Sulphur

58 **Strontianite**
Shape: Orthorhombic
Contains:
- Strontium
- Carbon
- Oxygen

59 **Sturmannite**
Shape: Hexagonal
Contains:
- Calcium
- Iron
- Aluminium
- Manganese
- Oxygen
- Hydrogen
- Boron
- Sulphur

60 **Sugilite**
Shape: Hexagonal
Contains:
- Potassium
- Sodium
- Lithium
- Iron
- Manganese
- Aluminium
- Silicon
- Oxygen

61 Topaz
Shape: Orthorhombic
Contains:
    Aluminium   Fluorine
    Silicon   Hydrogen
    Oxygen

62 Tourmaline
Shape: Trigonal/Hexagonal
Contains:
    Sodium   Oxygen
    Magnesium   Silicon
    Iron   Hydrogen
    Lithium   Fluorine
    Manganese   Aluminium
    Boron

63 Uvarovite
Shape: Cubic
Contains:
    Calcium
    Chromium
    Silicon
    Oxygen

64 Vanadinite
Shape: Trigonal/Hexagonal
Contains:
    Lead
    Vanadium
    Oxygen
    Chlorine

65 Willemite
Shape: Trigonal/Hexagonal
Contains:
    Zinc
    Silicon
    Oxygen

66 **Wulfenite**
Shape: Tetragonal
Contains:
Lead
Molybdenum
Oxygen

67 **Xenotime**
Shape: Tetragonal
Contains:
Yttrium
Phosphorus
Oxygen

68 **Zincite**
Shape: Hexagonal
Contains:
Zinc
Manganese
Oxygen

69 **Zircon**
Shape: Tetragonal
Contains:
Zirconium
Silicon
Oxygen

APPENDIX 3 – Table of Elements

Many attempts were made by Chemists during the nineteenth century to organise a table which positioned the elements in order of their atomic size and which also showed regular repeating patterns in their properties or behaviour. The most successful attempt was made by a Russian, Dimitri Mendeléev in 1869. The basis for the modern Periodic Table originates from Mendeléev's work. To make referencing easier, information provided here shows the elements in alphabetical order rather than by atomic weight.

Element Name	Element Abbreviation	Comments
Actinium	Ac	Actinide Series; Rare Earth
Aluminium	Al	Other Metal; Native
Americium	Am	Actinide Series; Transuranic; Rare Earth
Antimony	Sb	Metalloid; Native
Argon	Ar	Noble Gas
Arsenic	As	Metalloid; Native
Astatine	At	Halogen
Barium	Ba	Alkaline Earth Metal
Berkelium	Bk	Actinide Series; Transuranic; Rare Earth
Beryllium	Be	Alkaline Earth Metal
Bismuth	Bi	Other Metal; Native
Bohrium	Bh	Transition Metal; Transuranic (was Neilsborium)
Boron	B	Metalloid
Bromine	Br	Halogen
Cadmium	Cd	Transition Metal; Native
Caesium	Cs	Alkali-Metal
Calcium	Ca	Alkaline Earth Metal
Californium	Cf	Actinide Series; Transuranic; Rare Earth
Carbon	C	Mon-Metal; Native
Cerium	Ce	Lanthanide Series; Rare Earth
Chlorine	Cl	Halogen
Chromium	Cr	Transition Metal; Native
Cobalt	Co	Transition Metal
Copper	Cu	Transition Metal; Native
Curium	Cm	Actinide Series; Transuranic; Rare Earth
Dubnium	Db	Transition Metal; Transuranic (was Hahnium)
Dysprosium	Dy	Lanthanide Series; Rare Earth

Element Name	Element Abbreviation	Comments
Einsteinium	Es	Actinide Series; Transuranic; Rare Earth
Erbium	Er	Lanthanide Series; Rare Earth
Europium	Eu	Lanthanide Series; Rare Earth
Fermium	Fm	Actinide Series; Transuranic; Rare Earth
Fluorine	F	Halogen
Francium	Fr	Alkali-Metal
Gadolinium	Gd	Lanthanide Series; Rare Earth
Gallium	Ga	Other Metal
Germanium	Ge	Metalloid
Gold	Au	Transition Metal; Native
Hafnium	Hf	Transition Metal
Hassium	Hs	Transition Metal; Transuranic
Helium	He	Noble Gas
Holmium	Ho	Lanthanide Series; Rare Earth
Hydrogen	H	Non-Metal
Indium	In	Other Metal; Native
Iodine	I	Halogen
Iridium	Ir	Transition Metal; Native
Iron	Fe	Transition Metal
Krypton	Kr	Noble Gas
Lanthanum	La	Lanthanide Series; Rare Earth
Lawrencium	Lr	Actinide Series; Transuranic; Rare Earth
Lead	Pb	Other Metal; Native
Lithium	Li	Alkali-Metal
Lutetium	Lu	Lanthanide Series; Rare Earth
Magnesium	Mg	Alkaline Earth Metal
Manganese	Mn	Transition Metal
Meitnerium	Mt	Transition Metal; Transuranic
Mendelevium	Md	Actinide Series; Transuranic; Rare Earth
Mercury	Hg	Transition Metal; Native
Molybdenum	Mo	Transition Metal
Neodymium	Nd	Lanthanide Series; Rare Earth
Neon	Ne	Noble Gas
Neptunium	Np	Actinide Series; Transuranic; Rare Earth

Element Name	Element Abbreviation	Comments
Nickel	Ni	Transition Metal; Native
Niobium	Nb	Transition metal
Nitrogen	N	Non-Metal
Nobelium	No	Actinide Series; Transuranic; Rare Earth
Osmium	Os	Transition Metal
Oxygen	O	Non-Metal
Palladium	Pd	Transition Metal; Native
Phosphorus	P	Non-Metal
Platinum	Pt	Transition Metal; Native
Plutonium	Pu	Actinide Series; Transuranic; Rare Earth
Polonium	Po	Metalloid
Potassium	K	Alkali-Metal
Praseodymium	Pr	Lanthanide Series; Rare Earth
Promethium	Pm	Lanthanide Series; Rare Earth
Protactinium	Pa	Actinide Series; Rare Earth
Radium	Ra	Alkaline Earth Metal
Radon	Rn	Noble Gas
Rhenium	Re	Transition Metal; Native
Rhodium	Rh	Transition Metal; Native
Rubidium	Rb	Alkali-Metal
Ruthenium	Ru	Transition Metal
Rutherfordium	Rf	Transition Metal; Transuranic
Samarium	Sm	Lanthanide Series; Rare Earth
Scandium	Sc	Transition Metal; Rare Earth (sometimes)
Seaborgium	Sg	Transition Metal; Transuranic
Selenium	Se	Non-Metal; Native
Silicon	Si	Metalloid; Native
Silver	Ag	Transition Metal; Native
Sodium	Na	Alkali-Metal
Strontium	Sr	Alkaline Earth Metal
Sulphur	S	Non-Metal; Native
Tantalum	Ta	Transition Metal
Technetium	Tc	Transition Metal
Tellurium	Te	Metalloid; Native

Element Name	Element Abbreviation	Comments
Terbium	Tb	Lanthanide Series; Rare Earth
Thallium	Tl	Other Metal
Thorium	Th	Actinide Series; Rare Earth (sometimes)
Thulium	Tm	Lanthanide Series; Rare Earth
Tin	Sn	Other Metal; Native
Titanium	Ti	Transition Metal
Tungsten	W	Transition Metal
Ununbium	Uub	Transition Metal; Transuranic
Ununnilium	Uun	Transition Metal; Transuranic
Unununium	Uuu	Transition Metal; Transuranic
Uranium	U	Actinide Series; Rare Earth
Vanadium	V	Transition Metal
Xenon	Xe	Noble Gas
Ytterbium	Yb	Lanthanide Series; Rare Earth
Yttrium	Y	Transition Metal/Lanthanide; Rare Earth (usually)
Zinc	Zn	Transition Metal; Native
Zirconium	Zr	Transition Metal; Rare Earth (rarely)

APPENDIX 4 – Chemical Content of the 52 Gems/Minerals

Name	Crystal Type	Chemical Content	Colour
Almandine	Cubic	$Fe_3Al_2(SiO_4)_3$	Deep red, violet red, brownish-black
Analcime (Analcite)	Cubic	$NaAlSi_2O_6H_2O$	Greyish/Greenish
Andradite	Cubic	$Ca_3Fe_2(SiO_4)_3$	Green-yellow; orange-yellow; lime green to emerald green; brown-red; grey-green; dark green; brown, grey-black; black
Apatite	Hexagonal	$Ca_5(FCl)(PO_4)_3$	Blue, Colourless, Green, Pink, Violet, Yellow
Apophyllite	Tetragonal	$KFCa_4Si_8O_{12}(OH)_{16}$	Greenish Grey, Flesh, Red, White, Yellowish
Aquamarine	Hexagonal	$Be_3Al_2(SiO_3)_6$	Green
Barite (Baryte)	Orthorhombic	$BaSO_4$	White, Grey, Yellow, Blue, Red, Brown, Green
Boleite *	Tetragonal	$Pb(CuAg)Cl_2(OH)_2H_2O$	Deep Indigo Blue
Borax	Monoclinic	$Na_2B_4O_5(OH)_48H_2O$	White, Colourless
Bornite *	Cubic	Cu_5FeS_4	Iridescent Peacock Ore
Cassiterite (the principal ore of Tin)	Tetragonal	SnO_2	Black, Brown, Brownish-black, Colourless, Green, Grey
Cavansite	Orthorhombic	$Ca(V^{+4}O)Si_4O_{10} \cdot 4H_2O$	Blue
Chalcopyrite *	Tetragonal	$CuFeS_2$	Yellowish
Chromite	Cubic	$FeCr_2O_4$	Black, Brown
Chrysoprase	Trigonal/Hexagonal	SiO_2	Green
Cobaltite *	Cubic	$CoAsS$	Silvery White to Reddish
Covellite *	Hexagonal	CuS	Indigo, Midnight Blue
Cuprite *	Cubic	Cu_2O	Deep Red or Red Black
Cyanotrichite *	Orthorhombic	$Cu_4Al_2SO_4(OH)_{12} \cdot 2H_2O$	Blue
Diaboleite *	Tetragonal	$Pb_2CuCl_2(OH_4)$	Deep Blue
Dioptase *	Trigonal/Hexagonal	$CuSiO_2(OH)_2$	Green
Emerald	Hexagonal	$Be_3(AlCr)_2Si_6O_{18}$	Green
Fluorite (Fluorspar)	Cubic	CaF_2	Colourless, Blue, Brown, Green, Orange, Pink, Violet, Yellow

Name	Crystal Type	Chemical Content	Colour
Gaspeite	Trigonal/Hexagonal	$(NiMgFe)CO_3$	Green, Lime Green
Grossularite	Cubic	$Ca_3Al_2(SiO_4)_3$	Wine-Yellow, Brown, Green, Pale Green, Yellow, Red-Brown, Red, Orange, Colourless to White, Grey, Black
Hemimorphite	Orthorhombic	$Zn_4Si_2O_7(OH)_2 \cdot H_2O$	Pale Green to Green, Sky Blue, Various other shades of Blue, Colourless, White, Pale Yellow, Brown, Grey
Heulandite	Monoclinic	$(CaNaK)_{2-3}Al_3(AlSi)_2 S_{13}O_{36} 12H_2O$	White, Colourless, Grey, Yellow, Pink, Peach, Red, Green, Black, Brown
Idocrase (Vesuvianite)	Cubic	$Ca_{10}(MgFe)_2Al_4(SiO_4)_5(Si_2O_7)_2(OH)_4$	Blue, Brown, Green, Purple, Yellow
Jadeite	Monoclinic	$NaAl(SiO_3)_2$	Green, Blue, Brown, Black, Lilac, Mauve, Orange, Pink, Red, Violet
Labradorite	Trigonal/Hexagonal	$(CaNa)(SiAl)_4O_8$	Blue
Lazurite	Cubic	$(NaCa)_8(AlSiO_4)_6(SSiO_4Cl_2)$	Blue
Mimetite *	Hexagonal	$Pb_5(AsO_4)_3Cl$	Colourless, Brown, Grey, Green, Orange, Yellow, White
Mottramite *	Orthorhombic	$PbCu(VO_4)OH$	Green to Black
Nuummit	Cubic	Mixture of Anthophyllite $[(MgFe^{2+})_7Si_8O_{22}(OH)_2]$ and Gedrite $[(MgFe)_3Al_2(AlSi)_8O_{22}(OH)_2]$	Black with iridescent colour range of Gold, Copper, Purple, Silver, Green, Green-Blue and Blue
Pyrite	Cubic	FeS_2	Brass Yellow
Rhodonite	Triclinic	$(Mn^{+2}Fe^{+2}MgCa)SiO_3$	Pink to Rose Red, Brown, Red to Black
Rutile	Tetragonal	TiO_2	Black, Blue, Brown, Red, Violet, Yellow
Scapolite	Tetragonal	$Ca_4(SiAl)_{12}O_{24}(CO_3SO_4)$ and $(AlSi)_{12}O_{24}Cl$	Greenish, Pink, Violet, Yellow
Selenite	Monoclinic	$CaSO_4 2H_2O$	Whitish, Greyish, Greenish, Yellowish, Brownish, Reddish
Smithsonite	Trigonal/Hexagonal	$ZnCO_3$	White, Grey, Yellow, Green, Blue, Brown, Pink, Purple
Sodalite	Cubic	$Na_4Al_3(SiO_4)_3Cl$	Blue, Red, Green, Grey
Sphalerite	Cubic	$(ZnFe)S$	Green, Red, Yellow, White

Name	Crystal Type	Chemical Content	Colour
Strontianite	Orthorhombic	$SrCO_3$	White, Grey, Yellow, Brown, Green, Red
Sugilite	Hexagonal	$KNa_2Li_3(FeMnAl)_2Si_{12}O_{30}$	Pink, Violet, Purple
Topaz	Orthorhombic	$Al_2SiO_4(FOH)_2$	White, Colourless, Grey, Yellow, Orange, Brown, Bluish, Greenish, Purple, Pink
Tourmaline - Green	Trigonal/ Hexagonal	$Na(MgFeLiMnAl)_3Al_6(BO_3)_3Si_6O_{18}(OHF)_4$ The chemical content varies between each colour	Deep Pink to Red Orange, Green, Lavender, Black, Multi-coloured
Uvarovite	Cubic	$Ca_3Cr_2(SiO_4)_3$	Green
Vanadinite *	Trigonal/ Hexagonal	$Pb_5(VO_4)_3Cl$	Bright Red-Orange, Brownish-Red, Brown, Yellow
Wulfenite *	Tetragonal	$PbMoO_4$	Orange, Yellow, Brown, Grey, Greenish-Brown
Xenotime(y) *	Tetragonal	YPO_4	Yellowish-Brown to Reddish-Brown
Zincite	Hexagonal	$(ZnMn)O$	Orange, Red, Yellow
Zircon	Tetragonal	$ZrSiO_4$	Blue, Gold, Red, Green, Yellow, Colourless

* Minerals marked with an asterisk (*) contain toxic minerals and reference should be made to tincture preparation using the Indirect Method.

The books sourced for all this information are:

Published	Book	Publisher
2008	*Love Is In The Earth*, Melody	Earth-Love Publishing House, Colorado, USA
2008	*Gemstones*, Arthur Thomas	New Holland Publishers, UK.
2000	*Rocks & Minerals*, Chris Pellant	Dorling, Kindersley
2000	*Gemstones & Minerals of Australia*, Sutherland & Webb	Reed New Holland, Australia
2009	*Gemstones of the World,* Walter Schumann	Sterling Publishing Inc, New York
2003	*The Crystal Bible, Vol 1*, Judy Hall	Godsfield Press, UK

APPENDIX 5 – Chemical Content - Additional Hexagonal & Trigonal/Hexagonal Gems/Minerals

Name	Crystal Type	Chemical Content	Colour
Amethyst	Trigonal/Hexagonal	SiO_2	Deep Purple through to Pale Lavender
Aqua Aura Quartz (Enhanced with 24 carat gold)	Trigonal/Hexagonal	SiO_2Au	Blue-Green
Bismuth * (Treated sample)	Trigonal/Hexagonal	Bi	Silver-White, with Reddish-Pink hue
Chabazite	Trigonal/Hexagonal	$CaAl_2Si_4O_{12}6H_2O$	Whitish, Yellowish, Pinkish, Reddish, Greenish, Colourless
Citrine	Trigonal/Hexagonal	SiO_2	Yellow, Golden Brown, Burnt Amber
Cobalt Aura Quartz/ Siberian Blue Quartz (Enhanced with Cobalt)	Trigonal/Hexagonal	SiO_2Co	Deep Blue
Eudialyte *	Trigonal/Hexagonal	$Na_4(CaCe)_2Fe^{+2}Mn^{+2}Y) ZrSi_8O_{22})(OH,Cl)_2$	Pink, Rose, Red, Yellow-Brown, Brown, Yellow, Violet, Green, Red, Brown
Flame Aura Quartz (Enhanced with Titanium [Hexagonal] & Niobium [Cubic])	Trigonal/Hexagonal	SiO_2	Gold, Violet
Gyrolite	Trigonal/Hexagonal	$NaCa_{16}(Si_{23}Al)O_{60}(OH)_5 15H_2O$	White, Green (rarely), Brown, Grey
Magnesite	Trigonal/Hexagonal	$Mg(CO_3)$	Colourless, White, Grey, Yellowish, Brown
Painite	Hexagonal	$CaZrAl_9(O_{15})(BO_3)$	Deep Red
Phenakite	Trigonal/Hexagonal	Be_2SiO_4	Colourless, Wine-Yellow, Pink
Pyromorphite *	Hexagonal	$Pb_5(PO_4)_3Cl$	Green, Yellow-Green, Yellow, Orange-Yellow, Brown, White, Colourless
Quartz/Silver Enhanced with Silver	Trigonal/Hexagonal	SiO_2Ag	Clear, Silvery

Name	Crystal Type	Chemical Content	Colour
Rhodochrosite	Trigonal/Hexagonal	$MnCO_3$	Reddish-Pink, Raspberry Red, Pink
Sturmanite	Hexagonal	$Ca_6(Fe^{3+}AlMn^{2+})_2[(OH)_{12}/B(OH)_4/(SO_4)_2 \cdot 25H_2O$	Yellowish, Orange-Brown, Yellow-Green
Willemite	Trigonal/Hexagonal	Zn_2SiO_4	White, Colourless, Grey, Green, Yellow, Brown, Reddish, Pink, Black

* Minerals marked with an asterisk (*) contain toxic minerals and reference should be made to tincture preparation using the Indirect Method.

The books sourced for all this information are:

Published	Book	Publisher
2008	*Love Is In The Earth*, Melody	Earth-Love Publishing House, Colorado, USA
2008	*Gemstones*, Arthur Thomas	New Holland Publishers, UK.
2000	*Rocks & Minerals*, Chris Pellant	Dorling, Kindersley
2000	*Gemstones & Minerals of Australia*, Sutherland & Webb	Reed New Holland, Australia
2009	*Gemstones of the World*, Walter Schumann	Sterling Publishing Inc, New York
2003	*The Crystal Bible, Vol 1*, Judy Hall	Godsfield Press, UK

APPENDIX 6 – Where to find the Minerals

Mineral/Element	Chemical Symbol	Contained in the following crystals from the 69 selected:	Crystal Type
Calcium	Ca	Andradite	Cubic
		Apatite	Hexagonal
		Apophyllite	Tetragonal
		Cavansite	Orthorhombic
		Chabazite	Trigonal/Hexagonal
		Eudialyte	Trigonal/Hexagonal
		Fluorite	Cubic
		Grossularite	Cubic
		Gyrolite	Trigonal/Hexagonal
		Heulandite	Monoclinic
		Idocrase	Cubic
		Labradorite	Trigonal/Hexagonal
		Lazurite	Cubic
		Painite	Hexagonal
		Rhodonite	Triclinic
		Scapolite	Tetragonal
		Selenite	Monoclinic
		Sturmanite	Hexagonal
		Uvarovite	Cubic
Chloride	Cl	Boleite	Tetragonal
		Diaboleite	Tetragonal
		Eudialyte	Trigonal/Hexagonal
		Pyromorphite	Hexagonal
		Scapolite	Tetragonal
		Sodalite	Cubic
		Vanadinite	Trigonal/Hexagonal
Chromium	Cr	Chromite	Cubic
		Uvarovite	Cubic
Cobalt	Co	Cobalt Aura Quartz	Trigonal/Hexagonal
		Cobaltite	Cubic
		Rhodochrosite	Trigonal/Hexagonal

Mineral/Element	Chemical Symbol	Contained in the following crystals from the 69 selected:	Crystal Type
Copper	Cu	Boleite	Tetragonal
		Bornite	Cubic
		Chalcopyrite	Tetragonal
		Covellite	Hexagonal
		Cuprite	Cubic
		Cyanotrichite	Orthorhombic
		Diaboleite	Tetragonal
		Dioptase	Trigonal/Hexagonal
		Mottramite	Orthorhombic
Fluoride	F	Apatite	Hexagonal
		Apophyllite	Tetragonal
		Fluorite (Fluorspar)	Cubic
		Topaz	Orthorhombic
		Tourmaline	Trigonal/Hexagonal
Iron	Fe	Almandine	Cubic
		Andradite	Cubic
		Bornite	Cubic
		Chalcopyrite	Tetragonal
		Chromite	Cubic
		Eudialyte	Trigonal/Hexagonal
		Idocrase (Vesuvianite)	Cubic
		Nuummit	Cubic
		Pyrite	Cubic
		Rhodonite	Triclinic
		Sphalerite	Cubic
		Sturmanite	Hexagonal
		Sugilite	Hexagonal
		Tourmaline	Trigonal/Hexagonal
Magnesium	Mg	Idocrase (Vesuvianite)	Cubic
		Magnesite	Trigonal/Hexagonal
		Nuummit	Cubic
		Rhodonite	Triclinic
		Tourmaline	Trigonal/Hexagonal
Manganese	Mn	Eudialyte	Trigonal/Hexagonal
		Rhodochrosite	Trigonal/Hexagonal

Mineral/Element	Chemical Symbol	Contained in the following crystals from the 69 selected:	Crystal Type
		Rhodonite	Triclinic
		Sturmanite	Hexagonal
		Sugilite	Hexagonal
		Zincite	Hexagonal
Molybdenum	Mo	Wulfenite	Tetragonal
Nickel	Ni	Gaspeite	Trigonal/Hexagonal
Phosphorous	P	Apatite	Hexagonal
		Pyromorphite	Hexagonal
Potassium	K	Apophyllite	Tetragonal
		Heulandite	Monoclinic
		Sugilite	Hexagonal
Selenium	Se	Selenite	Monoclinic
Silicon	Si	Almandine	Cubic
		Amethyst	Trigonal/Hexagonal
		Analcime (Analcite)	Cubic
		Andradite	Cubic
		Apophyllite	Tetragonal
		Aqua Aura Quartz	Trigonal/Hexagonal
		Aquamarine	Hexagonal
		Cavansite	Orthorhombic
		Chabazite	Trigonal/Hexagonal
		Chrysoprase	Trigonal/Hexagonal
		Citrine	Trigonal/Hexagonal
		Cobalt Aura Quartz	Trigonal/Hexagonal
		Dioptase	Trigonal/Hexagonal
		Emerald	Hexagonal
		Eudialyte	Trigonal/Hexagonal
		Flame Aura Quartz	Trigonal/Hexagonal
		Grossularite	Cubic
		Gyrolite	Trigonal/Hexagonal
		Hemimorphite	Orthorhombic
		Heulandite	Monoclinic
		Idocrase (Vesuvianite)	Cubic

Mineral/Element	Chemical Symbol	Contained in the following crystals from the 69 selected:	Crystal Type
		Jadeite	Monoclinic
		Labradorite	Trigonal/Hexagonal
		Lazurite	Cubic
		Nuummit	Cubic
		Phenakite	Trigonal/Hexagonal
		Quartz/Silver	Trigonal/Hexagonal
		Rhodonite	Triclinic
		Scapolite	Tetragonal
		Sodalite	Cubic
		Sugilite	Hexagonal
		Topaz	Orthorhombic
		Uvarovite	Cubic
		Willemite	Trigonal/Hexagonal
		Zircon	Tetragonal
Sodium	Na	Analcime (Analcite)	Cubic
		Borax	Monoclinic
		Eudialyte	Trigonal/Hexagonal
		Gyrolite	Trigonal/Hexagonal
		Heulandite	Monoclinic
		Jadeite	Monoclinic
		Labradorite	Trigonal/Hexagonal
		Lazurite	Cubic
		Sodalite	Cubic
		Sugilite	Hexagonal
		Tourmaline	Trigonal/Hexagonal
Strontium	Sr	Strontianite	Orthorhombic
Sulphur	S	Barite (Baryte)	Orthorhombic
		Bornite	Cubic
		Chalcopyrite	Tetragonal
		Cobaltite	Cubic
		Covellite	Hexagonal
		Cyanotrichite	Orthorhombic
		Lazurite	Cubic
		Selenite	Monoclinic
		Sphalerite	Cubic

Mineral/Element	Chemical Symbol	Contained in the following crystals from the 69 selected:	Crystal Type
		Sturmanite	Hexagonal
Tin	Sn	Cassiterite	Tetragonal
Vanadium	V	Cavansite	Orthorhombic
		Mottramite	Orthorhombic
		Vanadinite	Trigonal/Hexagonal
Zinc	Zn	Hemimorphite	Orthorhombic
		Smithsonite	Trigonal/Hexagonal
		Sphalerite	Cubic
		Willemite	Trigonal/Hexagonal
		Zincite	Hexagonal

REFERENCES

Arrien, A 1993, *The Four-fold way – walking the paths of the warrior, teacher healer and visionary*, Harper, San Francisco.

Bansal, HL & Bansal, RS 1985, *Magnetic cure for common diseases*, Orient Paperbacks, Delhi.

Bhattacharya, AK 1985, *Gem therapy*, Firma KLM Private Ltd, Calcutta .

Boericke, W 1987, *Materia Medica with repertory,* 9th edn, B Jain Publishers Pvt Ltd, New Delhi.

Bourgault, L 1997, *The American Indian secrets of crystal healing,* Quantum, England.

Cayce, EE, Richards DG & Schwartzer, GC 2007, *Mysteries of Atlantis* A.R.E. Press, Virginia

Church Of England, The, The Order for The Burial of the Dead, *The Book of common prayer,* William Clowes & Sons Ltd, London.

Cunningham, S © 2002, *Cunningham's encyclopedia of crystal, gem & metal magic,* 2nd edn, Llewellyn Publications, Minnesota.

Elkin, AP © 1994, *Aboriginal men of high degree: Initiation and sorcery in the world's oldest tradition,* Inner Traditions, Vermont.

Emoto, M 2004a, *The Hidden messages in water,* Beyond Words Publishing Inc, Oregon.

Emoto, M 2004b, *The Healing power of water,* Hay House Inc.

Emoto, M 2005, *The True power of water,* Beyond Words Publishing Inc, Oregon.

Farndon, J 2007, *The Illustrated encyclopedia of rocks of the world*, Anness Publishing Ltd, London.

Ghadiali, D P 2003, *Spectro-chrome metry encyclopedia,* 5th edn, Dinshah Health Society, New Jersey.

Gerber, R 1988, *Vibrational medicine*, Bear & Company, New Mexico.

Gurudas, 1986, *Gem elixirs and vibrational healing volume II,* Cassandra Press, California.

Haas, EM 1992, *Staying healthy with nutrition*, Celestial Arts Publishers, California.

Hall, J 2003, *The Crystal bible a definitive guide to crystals,* Godsfield Press, London.

Hall, J 2009, *The Crystal bible volume 2,* Godsfield, London.

Hall, MP © 1972, *Healing, the diving art,* The Philosophical Research Society Inc, Los Angeles.

Harner, M 1990, *The Way of the shaman,* Harper & Row, New York.

Hawkes, JW 2007, *Cell-level healing,* Allen & Unwin, Australia.

Heline, C 1977, *Sacred science of numbers,* 4th edn, New Age Press Inc. California.

Holzapfel, E, Crépon, P & Philippe, C Translated from French by Transcript. 1986 (in UK), *Magnet therapy*, Thorsons Publishing Group, New York.

Joseph, F 2006, *The Lost civilisation of Lemuria,* Bear & Company, Vermont.

Koch, MU 1993, *Laugh with health,* Revised Edition. Spectrum Access Pty Ltd, Australia.

Kronberger, H & Lattacher, S English translation A Dubsky 1995, *On the track of water's secret,* Australian Print Group, Victoria.

Lad, V 1985, *Ayurveda the science of self-healing,* 2nd edn, Lotus Press, Wisconsin.

Lawlor, R 1991, *Voices of the first day,* Inner Traditions International Ltd, Vermont.

Lawrence, SB 1989, *Behind numerology,* Newcastle Publishing Co, North Hollywood.

Lewis, RM Supervisor 1984, *The Universe of numbers,* Supreme Grand Lodge of AMORC Inc, California.

Lipton, B 2005, *The Biology of belief,* Elite Books, Santa Rosa, USA.

MacIvor V & LaForest, S 1979, *Vibrations. Healing through color, homeopathy and radionics,* Samuel Weiser Inc, New York.

Martlew, G 1994, *Electrolytes the spark of life,* Nature's Publishing Ltd, Florida.

Melody 2008, *Love is in the earth - the crystal & mineral encyclopedia,* Earth-Love Publishing House, Colorado.

Narby, J 1999, *The Cosmic serpent – DNA and the origins of knowledge,* Jeremy P Tarcher Putman, New York.

Oschman, JL 2000, *Energy medicine – the scientific basis,* Churchill Livingstone, Edinburgh.

Ostrander, S & Schroeder, L 1973, *PSI – Psychic discoveries behind the iron curtain,* Sphere Books Ltd, Great Britain.

Pellant, C 2000, *Rocks & minerals,* Dorling Kindersley Ltd, London.

Reichel-Dolmatoff, G 1971, *Amazonian cosmos – the sexual and religious symbolism of the Tukano Indians,* The University of Chicago Press, Chicago.

Schauberger, V, translated and edited by C Coats, 1997, *The Water wizard,* Gateway, Dublin.

Schauberger, V, translated and edited by C Coats, 2000, *The Fertile earth,* Gateway, Dublin.

Schneider, MS 1995, *A Beginner's guide to constructing the universe,* Harper Perennial, New York.

Scholten, J 1996, *Homoeopathy and the elements,* Stichting Alonnissos, The Netherlands.

Schumann, W 2009, *Gemstones of the world,* Revised & expanded 4th edn, Sterling, New York/London.

Simmons, R & Ahsian, N, 2005, *The Book of stones,* Heaven & Earth Publishing LLC, Vermont.

Skow, D & Walters, C 1991, *Mainline farming for century 21,* Acres USA, Texas.

Stephenson, M 2008 *The Sage age blending science with intuitive wisdom,* Nightengale Press, Wisconsin.

Strehlow, W & Hertzka, G 1988, *Hildegard of Bingen's medicine,* Bear & Company, New Mexico.

Sutherland, L & Webb, G 2000, *Gemstones & minerals of Australia,* New Holland, Sydney.

Suzuki, D with McConnell, A 1997, *The Sacred balance,* Allen & Unwin, Australia.

Thomas, A 2008, *Gemstones: Properties, identification and use,* New Holland Publishers (UK) Ltd, London.

Tompkins, P & Bird, C 1992, *Secrets of the soil,* Arkana, London.

Walters, C 2006, *Minerals for the genetic code,* Acres USA, Texas.

BIBLIOGRAPHY

Alexandersson, O 1990, *Living water,* Gateway Books, Dublin.

Andersen, A 2004, *Real medicine, real health.* Holographic Health Press, USA.

Batmanghelidj, F 2001, *Your Body's many cries for water,* 2nd edn, Global Health Solutions Inc, Vienna, USA.

Davis, AR & Bhattacharya, AK 1982, *Healing by magnets,* Revised Edition. Firma KLM Private Ltd, Calcutta.

Frawley, D 1990, *The Astrology of the seers,* Passage Press, Utah.

Harding, J 2007, *Crystals,* Cameron House, South Australia.

Hewitt, WW 1992, *Astrology for beginners,* Llewellyn Publications, Minnesota.

Jansky, RC 1987, *Astrology nutrition and health*, Whitford Press, Pennsylvania.

Martin & Pleasance 1991, *Schuessler Tissue salts handbook.* Martin & Pleasance Pty Ltd. Australia.

Memmler, RL, Cohen, BJ & Wood, DL 1996, *The Human body in health and disease,* 8th edn, Lippincott-Raven, Philadelphia.

Ridder-Patrick J 1990, *A Handbook of medical astrology,* Arkana - The Penguin Group, London.

Schauberger, V, translated and edited by C Coats, 1997, *Nature as teacher,* Gateway, Dublin.

Schroeder, HA © 1973, *The Trace elements and man,* Nature's Publishing Ltd, Florida.

Simmons, R 2009, *Stones of the new consciousness,* Heaven & Earth Publishing LLC, Vermont, & North Atlantic Books, California.

Stockley, Oxlade, Werheim (n.d.) *The Usborne illustrated dictionary of science,* ISBN: 086020989X.

Watts, DL 2006, *Trace elements and other essential nutrients,* 5th Writers B-L-O-C-K Ed. Trace Elements, 4501 Sunbelt Drive, Addison, Texas, 75001.

INDEX

A

acidity 6, 34
acupuncture 4
alkalinity 6, 33, 49
Almandine 39
Amethyst 39
Analcime 39
Analcite 39
Andradite 39
Apatite 39
Apophyllite 39
Aqua Aura Quartz 39
Aquamarine 40
Axes 21

B

Barite 40
Big Bang 7
Bismuth 40
Boleite 40
Borax 40
Bornite 40

C

calcium 33
Cassiterite 41
Cavansite 41
Cell Membrane 19
Centre of Symmetry 21
Chabazite 41
Chalcopyrite 41
chloride 33
chlorine 33
chrome 33
Chromite 41
Chrysoprase 41
Citrine 41
cleansing the mineral 51
cleansing water soluble crystals 51
clear quartz crystal 22
cobalt 34
Cobalt Aura Quartz 42
Cobaltite 42
Conglomerates 16
copper 34
cosmic rays 27
cosmic serpent 23, 26, 28
Covellite 42
Crystal Symmetry 21
Cubic 21, 24

Cuprite 42
Cyanotrichite 42

D

Diaboleite 42
Dioptase 42
DNA 23
DNA double helix 26
dosages of tinctures 52
Dreamtime 7

E

earthquakes 15
emerald 30
Emerald 42
Endoplasmic Reticulum 19
Energetic Medicines 4
energy body 4
energy medicine 4, 5, 9
Eudialyte 43

F

Flame Aura Quartz 43
flower essences 4
fluorine 34
Fluorite 43
Fluorspar 43
fluroride 34
formation of minerals 31

G

Gaspeite 43
gem essences 4, 5
Gem Essences 5, 9
gemstone tincture 27
gold 30, 39
Grossularite 43
Gyrolite 43

H

heavy metals 32
Hemimorphite 44
Heulandite 44
Hexagonal 21
hexagonal crystals 26
hexagonal shapes 22, 23
homoeopathy 4

I

Idocrase	44
Igneous rocks	15
immune system	28
indigenous peoples	26
iodine	34
iron	35

J

Jadeite	44

K

Kirlian photography	4

L

Labradorite	44
Lazurite	44
Lemuria	1, 26

M

magma	15
Magnesite	44
magnesium	35
magnesium phosphate	6
magnetic field	13
magnetised water	10, 13
manganese	35
medicine man	28
Metamorphic rocks	16
Mimetite	44
mineral content	5, 21
mineral content of humans	6
miracle water	11
Mitochondria	19
molybdenum	35
Monoclinic	21
moonlight method	50
Mottramite	44

N

Nanotubes	19
nickel	36
niobium	43
non-physical body	4
Nucleus	19
number nine	24
number three	24
Nuummit	45

O

Orthorhombic	21, 24

P

Painite	45
Peacock Rock	40
pH	6, 14, 36, 49
Phenakite	45
phosphorous	36
physical body	2, 4, 5, 48
pigment colours	25
Planes of Symmetry	21
potassium	36
potassium phosphate	6
power of numbers	25
prism	27
Pyrite	45
Pyromorphite	45
Pythagoras	23

Q

quartz crystal	28, 29
Quartz Silver	45

R

radiant energy	25
Rhodochrosite	45
Rhodonite	46
Rock Cycle	15
rock types	15
Rutile	46

S

sardonyx	30
Scapolite	46
Sedimentary rocks	16
Selenite	46
selenium	36
Siberian Blue Quartz	42
silicon	37
Smithsonite	46
Sodalite	46
sodium	37
Sphalerite	47
Strontianite	47
strontium	37
Sturmanite	47
Sugilite	47
sulphur	37
sunlight method	50
superphosphate	14

T

tectonic plates	15
Tetragonal	21
Tin	37

tinctures	50
titanium	43
Topaz	47
Tourmaline	47
Black	48
Brown	48
Dark Blue	48
Green	48
Light Blue	48
Multicoloured	48
Pink	48
Purple-Violet	48
Red	48
Yellow	48
toxic impurities	27
toxic metals	32
Triclinic	21
Trigonal	21
tsunamis	15

U

Uvarovite	48

V

Vanadinite	49
vanadium	38
Vesuvianite	44
vibration	4, 21, 50
volcanoes	15

W

Willemite	49
Wulfenite	49

X

Xenotime	49

Z

zinc	38
Zincite	49
Zircon	49